CW00644391

THE
CURIOUS
WORLD OF
SCIENCE

THE CURIOUS WORLD OF SCIENCE

A visual miscellany of
STORIES, THEORIES,
DISCOVERIES & CURIOSITIES
plucked from the scientific world

SIMON FLYNN

CONTENTS

INTRODUCTION

science, *n.*
5.b. Science.
In modern use, often treated as synonymous with 'Natural and Physical Science', and thus restricted to those branches of study that relate to the phenomena of the material universe and their laws …

Oxford English Dictionary

To help guide you towards particular types of information, each entry is marked with an icon identifying the subject area you will discover there.

Measurement & Classification

Theory & Discovery

Mathematics & Formulae

Biography & History

Lists & Curiosities

Culture & Humour

SCIENCE.
What does that word conjure up in your mind? Writing over two hundred years ago, the great German authors Friedrich Schiller and Johann Wolfgang von Goethe gave this rather neat (if slightly silly-sounding) description in their collection of poetic epigrams, *Xenien*:

> *To one, it's a high, heavenly goddess. To another it's a cow that provides them with butter.*

You might want to read that again. What they appear to be saying is that science can be viewed in terms of powerful ends (I'm thinking of the very best metaphorical butter here) and almost spiritual, aesthetic means. In Schiller's and Goethe's eyes, at least, people's responses tend to depend on which of these they focus. But why not consider both? Why not acknowledge that one reason science is so special is that both these aspects can be true?

There's more though. Science isn't just about laws, theories, formulae, processes and experiments. At heart it's a human activity. Without the incredible individuals who have occupied themselves with uncovering the workings of nature and applying them to our benefit, where would we be? That's the last rhetorical question, I promise.

This miscellany is intended to showcase just some of the many and varied facets of science, plenty of which are unashamedly and idiosyncratically human. It's people who provide great stories and heart-warming anecdotes. Partly for this reason, the story of Erasto Mpemba, an African schoolboy who didn't give up in his quest to understand something (and who has a physical effect named after him as a result), and the note of English naturalist Charles Darwin on the pros and cons of marriage are among my favourite entries in these pages.

Although this book is a miscellany, certain themes and ideas echo through its pages. One of these is mathematics. You may have noticed the '…' at the end of the definition of science given at the start of this introduction. That's because I cheated slightly – the definition actually continues, '… sometimes with implied exclusion of pure mathematics'. However, I side firmly with the 13th-century English philosopher Roger Bacon, who said, 'Mathematics is the door and the key to the sciences' and, as such, it's included. At least I'm honest.

A non-scientific realm that features rather heavily in the book is the arts, and particularly poetry. In contrast with science, artistic expression of any kind is typically felt to possess a power beyond the material. I'm not sure quite how fair a view of science this is. But I believe that an appreciation of both realms of human endeavour can only be for the good, given the great wonders each offers. So, as writers are often told, 'show, don't tell'. That has been the intention in those places in the book where the two cultures overlap, such as the entry 'The poets'

scientist' and the poem 'Letter from Caroline Herschel' (see pages 26 and 122) by Swedish-American author Siv Cedering.

And so, to the final big theme in *The Curious World of Science*. The splendid New Zealander short-story writer Katherine Mansfield once wrote, 'It is of immense importance to learn to laugh at ourselves' – and I couldn't agree more. From a smattering of groan-inducing jokes in some of the boxed fillers to moments of more delicate wit from practitioners of science in entries such as 'Ban Dihydrogen Monoxide' and 'In Chemic union', science's lighter side is very much on show because, as the French author Colette put it, 'Total absence of humour renders life impossible'.

Science has the power to enrich people's lives, both figuratively and practically. What follows is a celebration of it, warts and all. I hope you enjoy it.

P.S.
In case it has been a while since you last did, or read, any science, I've put together a brief back-to-school appendix (see page 214), which will hopefully remind you of some of the science you were taught at school and come in useful when reading this book. Don't worry, there won't be any exam at the end of it.

Simon Flynn

'Hymn to Science'

A PATIENT HEAD AND A CANDID HEART

It seems only fitting to open *The Curious World of Science* with a 'Hymn to Science', which first appeared in *The Gentleman's Magazine* in 1739, when its author, Mark Akenside, was only seventeen. It was around this time that Akenside, the son of a Newcastle butcher, switched from preparing for a life as a nonconformist clergyman to training in medicine. He soon became a member of the Medical Society in Edinburgh, eventually securing the position of physician to Queen Charlotte, wife of King George III, a little over twenty years later. Akenside wrote poetry throughout his life, including continuously revising his most famous work, *The Pleasures of the Imagination*, which the great English man of letters Dr Johnson described as 'an example of great felicity of genius'.

From the 'Hymn to Science'

Science! thou fair effusive ray
From the great source of mental Day,
Free, generous, and refin'd!
Descend with all thy treasures fraught,
Illumine each bewilder'd thought,
And bless my lab'ring mind.

But first with thy resistless light,
Disperse those phantoms from my sight,
Those mimic shades of thee;
The scholiast's learning, sophist's cant,
The visionary bigot's rant,
The monk's philosophy.

O! let thy powerful charms impart
The patient head, the candid heart,
Devoted to thy sway;
Which no weak passions e'er mislead,
Which still with dauntless steps proceed
Where Reason points the way.

Give me to learn such secret cause;
Let number's, figure's, motion's laws
Reveal'd before me stand;
These to great Nature's scenes apply,
And round the globe, and thro' the sky,
Disclose her working hand.

Next, to thy nobler search resign'd,
The busy, restless, human mind
Thro' ev'ry maze pursue;
Detect Perception where it lies,
Catch the ideas as they rise,
And all their changes view.

Say from what simple springs began
The vast ambitious thoughts of man,
Which range beyond controul,
Which seek Eternity to trace,
Dive thro' th' infinity of space,
And strain to grasp The Whole.

How Do They Do IT?

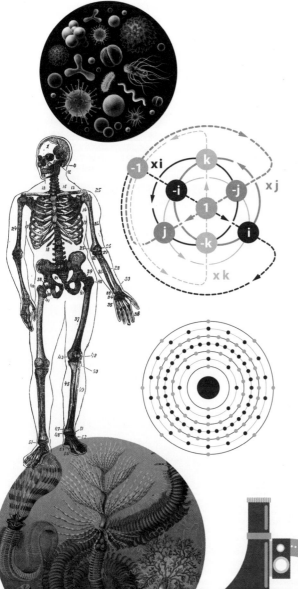

IN THE DARK AND ON THE TABLE

How do they do **IT**?

Astronomers do **IT** in the dark. **Mathematicians** do **IT** in numbers. **Biologists** do **IT** in the field. **Chemists** do **IT** periodically on the table. **Geologists** do **IT** in folded beds. **Palaeontologists** do **IT** in the dirt. **Computer scientists** do **IT** bit by bit. **Electrical engineers** do **IT** until it hertz. **Physicists** do **IT** with force. When **seismologists** do **IT,** the Earth shakes. **Zoologists** do **IT** with animals. **Quantum physicists** do **IT** uncertainly. **Polymer chemists** do **IT** in chains. **Cosmologists** do **IT** with a bang. **Theorists** do **IT** on paper. **Geneticists** do **IT** in their genes. **Statisticians** do **IT** with 99 per cent confidence. **Planetary scientists** do **IT** while gazing at Uranus. **Philosophers** only think about doing **IT**.

What is **IT**? Why, science of course. And shame on you, if you thought otherwise.

ADAPTED FROM JUPITER SCIENTIFIC (WWW.JUPITERSCIENTIFIC.ORG)

Congratulations, You've Won a …

TAKING ON A PROBABILITY PUZZLE

Originally proposed by Steve Selvin, an American statistician, in a letter to *The American Statistician* in 1975, and inspired by the US TV gameshow *Let's Make a Deal* hosted by Canadian broadcaster Monty Hall, this probability puzzle is appropriately named the Monty Hall problem.

> *He uses statistics as a drunken man uses lampposts – for support rather than for illumination.*
>
> **ANDREW LANG (1844–1912), SCOTTISH INTELLECTUAL AND WRITER**

You're on a game show and have got through to the final round. The grand prize is almost within your grasp. The host presents you with three doors, numbered 1–3. Behind one of these is a luxury car; behind the other two are goats. You choose a door. Before it's opened, the host, who knows what's behind each door, opens one of the others, revealing one of the goats. He then asks you if you want to change your mind from your original choice to the other unopened door. The question is whether this would be to your advantage.

Instinctively, we all feel that this would make absolutely no difference at all, but in fact the answer is 'yes', absolutely you should switch. If you change your choice, your odds of winning the car will become 2/3, whereas if you stick with your original choice, it'll be 1/3. Don't believe me? Here's how it works:

Let's assume you picked door 1. We can now show the possible arrangements of the prizes (if a goat is a prize).

So, if your initial choice of door has a goat behind it (if, say, you picked door 1 and there was a goat, meaning the set-up was either arrangement B or C), then switching your choice will definitely win you the car because Monty Hall would open the other door with a goat behind it. In two out of the three possible arrangements, switching choice would win you the car. Sticking with your choice would win you the car in only one of the three arrangements. Taking arrangement A as an example, if you picked door 1, then you'd win by sticking. But if you picked either door 2 or door 3, then you'd win by switching. You see?

A version of the problem was printed in *Parade* magazine in 1990 and prompted nearly 10,000 complaints that the magazine had got it wrong and that there was no benefit to the switch.

THE MONTY HALL PROBLEM

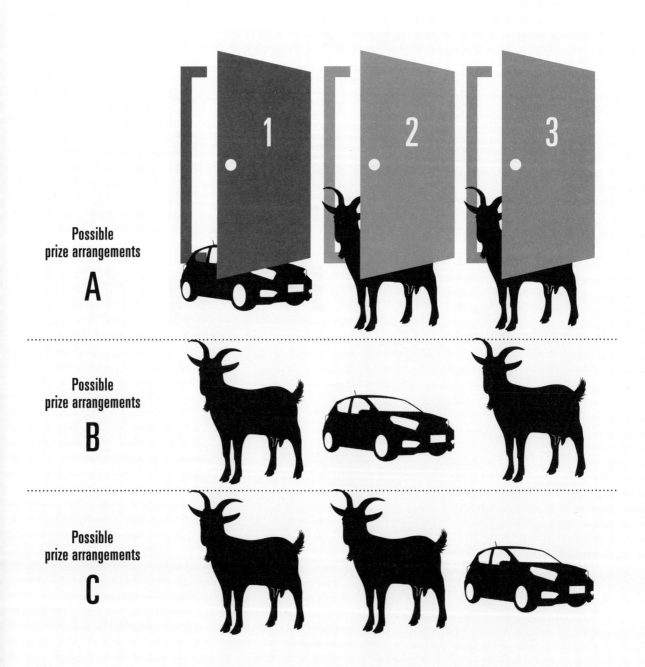

Possible prize arrangements
A

Possible prize arrangements
B

Possible prize arrangements
C

To Marry or
Not to Marry,
that Is the Question

DARWIN'S MARRIAGE DILEMMA

During the summer of 1838, English naturalist Charles Darwin, then aged 29, found himself in a bit of a quandary. The subject worrying him was whether he should marry or not, and, if so, when. To help him come to a decision, he wrote a note listing the pros and cons of having a wife, and, as would be expected from the man who would revolutionise evolutionary theory through the examination of argument and evidence, he really cut to the heart of the matter:

This Is the Question

Marry

Children — (if it Please God) — Constant companion, (& friend in old age) who will feel interested in one, — object to be beloved & played with.— — better than a dog anyhow. — Home, & someone to take care of house — Charms of music & female chit-chat. — These things good for one's health. — but terrible loss of time. —

My God, it is intolerable to think of spending one's whole life, like a neuter bee, working, working, & nothing after all. — No, no won't do. — Imagine living all one's day solitarily in smoky dirty London House.— Only picture to yourself a nice soft wife on a sofa with good fire, & books & music perhaps — Compare this vision with the dingy reality of Grt. Marlbro' St.

Not marry

Freedom to go where one liked — choice of Society & little of it. — Conversation of clever men at clubs — Not forced to visit relatives, & to bend in every trifle. — to have the expense & anxiety of children — perhaps quarelling — Loss of time. — cannot read in the Evenings — fatness & idleness — Anxiety & responsibility — less money for books &c — if many children forced to gain one's bread. — (But then it is very bad for one's health to work too much).

Perhaps my wife won't like London; then the sentence is banishment & degradation into indolent, idle fool —

Darwin's eventual conclusion?

Marry—Marry—Marry Q.E.D.

Darwin proposed to his cousin Emma Wedgwood on 11 November 1838, writing in his journal 'the day of days!'. They were married two and a half months later, on 29 January 1839.

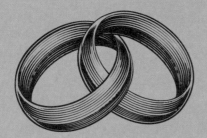

What's a Mole Worth?

THE STRANGE WORLD OF AVOGADRO'S CONSTANT

A mole can mean many things. It could be a small mammal that spends most of its time underground and causes consternation to gardeners around the world, a type of skin growth that's typically coloured, a spy who infiltrates the secret services of another country or, most importantly, here, a unit of measurement in chemistry. In Latin, *moles* means 'heap' or 'mass'.

To say you have a mole of a substance is to have 6.02×10^{23} units of its formula. For example, one mole of pure water contains 6.02×10^{23} H_2O molecules, and one mole of gold is made up of 6.02×10^{23} Au atoms. This curious number is known as Avogadro's constant or number and is named after the Italian lawyer Amadeo Avogadro (1776–1856). He determined that equal volumes of all gases have the same number of particles at the same temperature and pressure. At room temperature and pressure, one mole of any gas has a volume of $24,000 \text{cm}^3$.

To better appreciate just how big Avogadro's number is, one mole of water is roughly a small mouthful. So, if you were to count every molecule this amount contained, one per second, you would need the universe up to the here and now to have happened more than one million times (the age of the universe is approximately 13.8 billion years, which equates to roughly 4×10^{17} seconds).

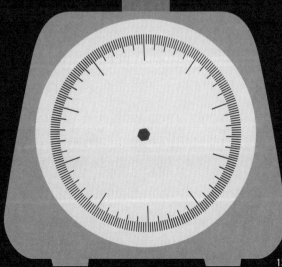

The website **what-if.xkcd.com** attempted to answer the question 'what would happen if you were to gather a mole of moles in one place?'. One of the things it determined was that collecting this many moles would result in an object a little larger than the Moon. I urge you to read their answer in full.

The True Measure of Things (Zooming In)

1×10^{-9}m
Diameter of
DNA helix

EVER DECREASING LENGTHS
A scale comparing the sizes of very
small things. Units that you might have
heard of are: 10^0m = one metre; 10^{-3}m =
millimetres (mm); 10^{-6}m = micrometres
(μm); 10^{-9}m = nanometres (nm).

1×10^{-15}m
Diameter of a proton

1×10^{-10}m
Diameter of a hydrogen atom

1×10^{-7}m
Size of Coronavirus

| 10^{-15}m | 10^{-14}m | 10^{-10}m | 10^{-9}m | 10^{-8}m | 10^{-7}m |

1.5×10^{-10}m
C–C bond length
in diamond

1×10^{-9}m
Length of a
glucose
molecule

$1-1.5 \times 10^{-8}$m
Thickness of a typical
cell membrane

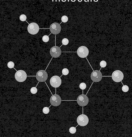

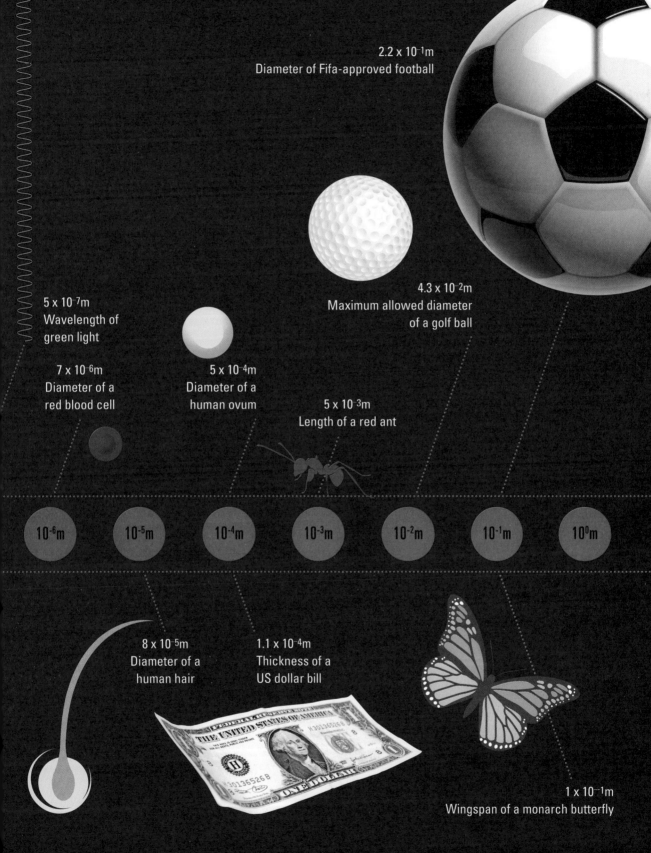

2.2×10^{-1}m
Diameter of Fifa-approved football

5×10^{-7}m
Wavelength of
green light

4.3×10^{-2}m
Maximum allowed diameter
of a golf ball

7×10^{-6}m
Diameter of a
red blood cell

5×10^{-4}m
Diameter of a
human ovum

5×10^{-3}m
Length of a red ant

10^{-6}m	10^{-5}m	10^{-4}m	10^{-3}m	10^{-2}m	10^{-1}m	10^{0}m

8×10^{-5}m
Diameter of a
human hair

1.1×10^{-4}m
Thickness of a
US dollar bill

1×10^{-1}m
Wingspan of a monarch butterfly

As Easy as
Al, Be, Cs

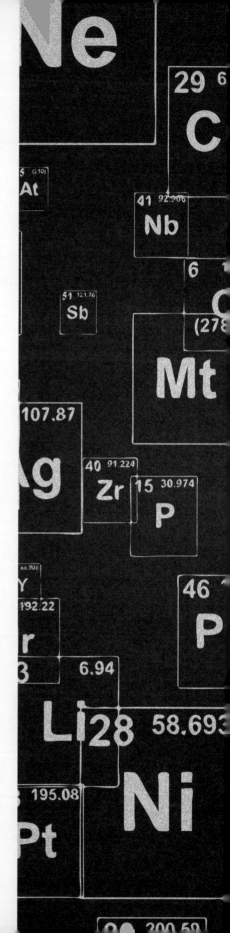

SINGING THE PERIODIC TABLE

There are a great many popular songs inspired by love, loss and, oh, did I mention love? But only one song can claim to have been inspired by an object of adoration quite as unusual as the periodic table of elements.

Tom Lehrer, born in New York in 1928, is a retired satirical songwriter who released a number of very successful albums in the 1950s and 60s. He also happened to be an eminent mathematician and lecturer. One of his songs, 'The Elements', is a version of the then 102-element periodic table to the tune of the 'Major-General's Song' from the comic opera *The Pirates of Penzance* by Gilbert and Sullivan. Versions by various people can be found online, including one by Daniel Radcliffe, the actor who played Harry Potter. However, the best undoubtedly remains the live version sung by Lehrer himself.

Tom Lehrer

*Now if I may digress momentarily from the mainstream of this evening's symposium,
I'd like to sing a song, which is completely pointless but it's something I picked up during
my career as a scientist. This may prove useful to some of you some day, perhaps in a
somewhat bizarre set of circumstances. It's simply the names of the chemical elements
set to a possibly recognisable tune.*

There's antimony, arsenic, aluminum, selenium
And hydrogen and oxygen and nitrogen and rhenium
And nickel, neodymium, neptunium, germanium
And iron, americium, ruthenium, uranium
Europium, zirconium, lutetium, vanadium
And lanthanum and osmium and astatine and radium
And gold and protactinium and indium and gallium
And iodine and thorium and thulium and thallium

There's yttrium, ytterbium, actinium, rubidium
And boron, gadolinium, niobium, iridium
And strontium and silicon and silver and samarium
And bismuth, bromine, lithium, beryllium, and barium

*(Isn't that interesting? [Audience laughs] I knew you would. I hope you're all taking
notes, because there's going to be a short quiz next period.)*

There's holmium and helium and hafnium and erbium
And phosphorus and francium and fluorine and terbium
And manganese and mercury, molybdenum, magnesium
Dysprosium and scandium and cerium and caesium
And lead, praseodymium, and platinum, plutonium, palladium, promethium,
potassium, polonium
And tantalum, technetium, titanium, tellurium
And cadmium and calcium and chromium and curium

There's sulphur, californium, and fermium, berkelium
And also mendelevium, einsteinium, nobelium
And argon, krypton, neon, radon, xenon, zinc, and rhodium
And chlorine, carbon, cobalt, copper, tungsten, tin, and sodium

These are the only ones of which the news has come to Ha'vard
And there may be many others but they haven't been discavard

A Room Full of Monkeys

THE SQUARE OF SCIENTIFIC DELIGHTS

London's Leicester Square has long been regarded as a centre of entertainment. But what might come as a surprise is that, of the various diversions found there through history, quite a few have been scientific in nature.

On 4 February 1775, *The Morning Post and Daily Advertiser* ran the following advert on its front page:

> ### *Museum, Leicester House, Feb. 3, 1775*
>
> *Mr. Lever's Museum of Natural and other Curiosities, consisting of beasts, birds, fishes, corals, shells, fossils extraneous and native, as well as many miscellaneous articles in high preservation, will be opened on Monday 13th of February, for the inspection of the public. … As Mr. Lever has in his collection some very curious monkies and monsters, which might disgust the Ladies, a separate room is appropriated for their exhibition, and the examination of those only who chuse it.*

The Holophusicon ('embracing all of nature'), as it was sometimes called, was located at Leicester House on the northern side of Leicester Square. It held the natural history collection of Sir Ashton Lever, which visitors could view for half a guinea. The collection, of around 27,000 objects, included a hippopotamus, an elephant, hummingbirds, pelicans, peacocks, bats, lizards and scorpions as well as many artefacts picked up during the voyages of English explorer Captain James Cook. Writing in 1778 to her cousin Frances (author of the best-selling novels *Evelina* and *Cecilia*), Susan Burney described the infamous room that might 'disgust the ladies' as

> … full of monkeys – one of which presents the company
> with an Italian Song – another is reading a book
> – another, the most horrid of all, is put in the attitude
> of Venus de Medicis, and is scarce fit to be look'd at.

SCIENTIFIC DELIGHTS

An illustration of Leicester Square,
London, from 1880.

The collection had initially been shown at Lever's country house, Alkrington Hall, near Manchester, before moving to London because it wasn't making enough money to sate Lever's addiction for collecting. Unfortunately, the London museum couldn't sustain itself either, despite being extremely popular and visited by King George III and the Prince of Wales. Lever ended up having to sell his great miscellany by lottery in 1786 (the British Museum had first refusal but sadly declined) and it ended up fragmenting. Never quite able to cope with this loss, Lever committed suicide less than two years later.

In 1783, while the Holophusicon was still going strong, the Scottish anatomist John Hunter began renting a large house in Leicester Square, where he ran a medical school and museum. Hunter was another prodigious collector, and people now had the opportunity to view skeletons of kangaroos from one of Cook's voyages, as well as the skeleton of Charles Byrne, an Irishman measuring around 2.31m (7ft 7in). The acquisition of the latter, which cost Hunter the equivalent of around £20,000 in today's money, is at the centre of the 1998 novel *The Giant, O'Brien*, by English writer Hilary Mantel. Unlike Lever's collection, Hunter's was thankfully bought by the British government after the anatomist's death and now forms the core of the Hunterian Museum at the Royal College of Surgeons in London.

Seventy years later, Leicester Square witnessed the creation of its most ambitious scientific establishment yet. Sadly, it also proved to be its last. The Royal Panopticon of Science and Art, which dominated the eastern side of the square, was built in the Moorish style with a 'towering minaret', 'lofty dome' and 'abundant use of chromatic decoration'. Its aim was 'to exhibit and illustrate, in a popular form, discoveries in science and art'. The establishment opened its doors to the general public in 1854 to considerable fanfare. Inside, no expense had been spared. Placed beneath the dome was a fountain, the central jet of which shot up almost 30m (100ft). At the entrance of the western gallery was the largest organ in England at the time, and a lift (referred to as an 'ascending carriage'), which could carry eight persons at a time, transported visitors to the photography gallery. Among the displays were an aurora borealis apparatus, which enabled the creation of artificial 'northern lights', a 'crystal cistern for diving' and a gas cooker (a reviewer lamented it not being put to better use by cooking his dinner). In the basement were lecture theatres where demonstrations were regularly given. It was, *The Morning Post* enthused, 'the most magnificent temple erected for the purposes of science'. It may well have been, but few of the public paid homage. Despite its royal charter, Queen Victoria failed to grace the 'temple' and two years later the venture was declared bankrupt.

EDUCATED APES
Letzte Vorstellung (Last Performance) by Prague-born artist Gabriel von Max (1840–1915).

A Call for Concision

VARIATIONS ON A THEME OF

Entia non sunt multiplicanda praeter necessitatem.

No more things should be presumed to exist than are absolutely necessary.

The above is the standard version of Ockham's Razor, named after William of Ockham, a 14th-century English Franciscan monk and philosopher. It's also known as the 'principle of parsimony' and has often been held up as a useful rule of thumb with which to judge the relative merits of two competing scientific theories that predict, or account for, the same experimental results. The razor is generally understood to mean the application of a preference to the simpler theory.

However, the point of the razor isn't primarily about simplest always being best. Instead, it's a call to include only what's necessary (the two often go hand in hand, but it's the latter that's the driver). The problem with this principle is that it isn't always clear in science what's necessary – or even whether all the relevant information to enable a judgment is available. As such, this explains why the principle isn't embraced universally.

William of Ockham appears to have written various versions of the principle, and many other thinkers have expressed their own versions.

William of Ockham

> *God made the integers; all the rest is the work of man.*
> **LEOPOLD KRONECKER (1823–1891), GERMAN MATHEMATICIAN**

ARISTOTLE:

'Nature operates in the shortest way possible.'

'If the consequences are the same, it is always better to assume the more limited antecedent.'

WILLIAM OF OCKHAM:

'Plurality is not to be posited without necessity.'

'No plurality should be assumed unless it can be proved by reason, or by experience, or by some infallible authority.'

'It is futile to do with more things that which can be done with fewer.'

ISAAC NEWTON:

'We are to admit no more causes of natural things than such as are both true and sufficient to explain their appearances. Therefore, to the same natural effects we must, so far as possible, assign the same causes.'

BERTRAND RUSSELL:

'Whenever possible, substitute constructions out of known entities for inferences to unknown entities.'

JOHANNES KEPLER:

'Nature uses as little as possible of anything.'

ERNST MACH:

'Scientists must use the simplest means of arriving at their results and exclude everything not perceived by the senses.'

ALBERT EINSTEIN:

'Everything should be made as simple as possible, but not simpler.'

Perhaps the German-American architect Ludwig Mies van der Rohe put it most succinctly of all when he said simply:

'Less is more'.

Name that Number

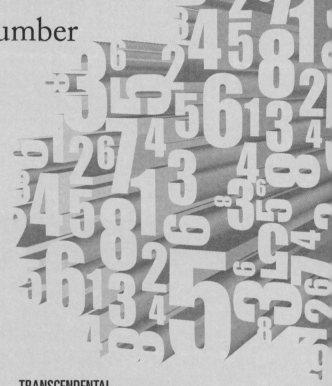

FROM NATURAL TO TRANSCENDENTAL AND BEYOND

You may realise this already, but there exist many different types of numbers. As such, it's possible to classify numbers much as we do plants and animals. The following is a quick number classification primer.

NATURAL

These are also known as counting numbers, and are 1, 2, 3 …

INTEGERS

These are like natural numbers but augmented by zero and negative numbers. They are always whole – for example,
–3, –2, –1, 0, 1, 2, 3 …

RATIONAL

Any number that's an integer or a ratio of integers – for example, $-2, -\frac{33}{40}, \frac{1}{2}, 7$ …

IRRATIONAL

Any number that cannot be represented by an integer or ratio of integers – for example, $\sqrt{5}$, e or π. Their decimals are infinite and don't repeat – for example, $\sqrt{5}$ (the square root of 5) is 2.2360679774 …

TRANSCENDENTAL

These are irrational numbers that aren't roots of any algebraic equation with rational coefficients. e is a transcendental number, but wouldn't be if, say, $5e^2 + 2e + 20 = 0$ (5 and 2 in this equation are rational coefficients). It's actually very difficult to prove a number is transcendental, which partly explains why e and π weren't shown to be so until 1873 and 1882 respectively.

REAL

Any number that's rational or irrational, so every type of number mentioned in this entry so far is real. The symbol for the set of all real numbers is R.

COMPLEX

These numbers are of the form $a + bi$, where a and b are any real number and $i^2 = -1$. a is said to be the real part and b the imaginary part of the number. Examples of imaginary numbers include i ($\sqrt{-1}$) and $7i$ ($\sqrt{-49}$).

A Thing of Mathematical Beauty

THE 'GREATEST EQUATION EVER'

Leonhard Euler (pronounced 'oiler') is one of the most renowned mathematicians ever to have lived. Born in Basel, Switzerland, in 1707, he wrote close to 900 papers and books despite losing the sight in his right eye in his early twenties and going completely blind when he was sixty-four.

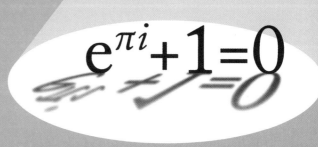

$$e^{\pi i} + 1 = 0$$

Euler's achievements are numerous – he solved the mathematical problem known as the Seven Bridges of Königsberg (leading to the development of graph theory), investigated the mathematics of music and proved Fermat's little theorem. He was also responsible for introducing, or popularising, many pieces of mathematical notation still used today, including e, π and i ($\sqrt{-1}$).

In the course of his work, Euler derived the equation shown above. Known as Euler's identity, it links together what the Israeli mathematician Eli Maor has called the five most important constants in mathematics: e, i, π, 1 and 0. The equation was described by American physicist Richard Feynman as 'the most remarkable formula in mathematics', voted the most beautiful theorem in mathematics by readers of the journal *The Mathematical Intelligencer* and emerged as the joint 'greatest equation ever' (along with Maxwell's equations of electromagnetism) in a poll in *Physics World*. It's wonderfully captured in the limerick:

e raised to the pi times i,
And plus 1 leaves you nought
    but a sigh.
This fact amazed Euler
That genius toiler,
And still gives us pause,
    bye the bye.

'In Chemic union'

THE SUBLIME ART OF SCIENTIFIC WOOING

English writer Constance Naden (1858–1889), an admirer of the philosopher and sociologist Herbert Spencer who coined the phrase 'survival of the fittest', published two volumes of poetry during her relatively brief life. Writing in a magazine, *The Speaker*, in 1890, the politician William Gladstone named her as one of the best 'poetesses' of that century in a list that also included Emily Brontë, Elizabeth Barrett Browning and Christina Rosetti.

Constance Naden

Naden was an unusual poet for the Victorian age. For a start, she was a woman who was interested in and studied science – an area assumed by many to favour the male mind. Her poetry often touched upon the realms of science, and nowhere is this better demonstrated than in her witty series of four poems 'Evolutional Erotics'. In these, she presents the relationships of four couples, using the sciences to great effect in her metaphor and imagery. Below is an extract from the first in the series, which describes how a young scientist's passion for his studies is transferred to an, at best, unsuspecting and, at worst, uninterested Mary Maud Trevylyan.

A modern poet has characterised the personality of art and the impersonality of science as follows: art is I, science is we.

CLAUDE BERNARD (1813–1878), FRENCH PHYSIOLOGIST

From 'Scientific Wooing'

I WAS a youth of studious mind,
Fair Science was my mistress kind,
And held me with attraction chemic;
No germs of Love attacked my heart,
Secured as by Pasteurian art
Against that fatal epidemic.

[...]

Alas! that yearnings so sublime
Should all be blasted in their prime
By hazel eyes and lips vermilion!
Ye gods! restore the halcyon days
While yet I walked in Wisdom's ways,
And knew not Mary Maud Trevylyan!

[...]

I covet not her golden dower—
Yet surely Love's attractive power
Directly as the mass must vary—
But ah! inversely as the square
Of distance! shall I ever dare
To cross the gulf, and gain my Mary?

[...]

Bright fancy! can I fail to please
If with similitudes like these
I lure the maid to sweet communion?
My suit, with Optics well begun,

By Magnetism shall be won,
And closed at last in Chemic union!

At this I'll aim, for this I'll toil,
And this I'll reach—I will, by Boyle,
By Avogadro, and by Davy!
When every science lends a trope
To feed my love, to fire my hope,
Her maiden pride must cry 'Peccavi!'

I'll sing a deep Darwinian lay
Of little birds with plumage gay,
Who solved by courtship Life's enigma;
I'll teach her how the wild-flowers love,
And why the trembling stamens move,
And how the anthers kiss the stigma.

Or Mathematically true
With rigorous Logic will I woo,
And not a word I'll say at random;
Till urged by Syllogistic stress,
She falter forth a tearful 'Yes,'
A sweet 'Quod erat demonstrandum!'

It Takes Two

BINARY BASICS

The number system we use in everyday life is base ten. There are many theories as to why this is the case, the most popular being that we have ten digits on our hands. Various cultures have adopted other counting systems, such as the ancient Babylonians, who used base sixty, giving us sixty minutes in an hour.

Today, base ten dominates, but there is another system, binary, that uses base two. This means that only two symbols, typically 0 and 1, are needed to express any value. The upper table shows how some familiar base ten numbers look in binary.

The lower table shows a variety of other base ten numbers, up to a maximum of 127, using binary numeration. By employing the binary system, it's possible to count to 1,023 using just your two hands, if a finger or thumb up stands for 1 and a finger or thumb down represents 0 (so, all ten digits up would represent 1111111111 = 1,023 in base ten).

Binary is particularly useful for computers because the digits 1 and 0 can be expressed in a variety of simple paired absolutes, such as on/off, yes/no or electric potential/no electric potential. In the book *The Hitchhikers Guide to the Galaxy*, by English writer Douglas Adams, the supercomputer Deep Thought gave the answer to the meaning of life as 42. However, the machine would probably have found it more natural to express the answer as 101010, which has a rather nice symmetry to it, don't you think? Binary has also given rise to one of the very best maths jokes:

There are only 10 types of people in the world: those who understand binary, and those who don't.

OK, I'll get my coat.

The base ten numbers	2^6 or 64	2^5 or 32	2^4 or 16	2^3 or 8	2^2 or 4	2^1 or 2	2^0 or 1
1							1
2						1	0
3 (=2+1)						1	1
4					1	0	0
5 (4+1)					1	0	1
10 (=8+2)				1	0	1	0
100 (=64+32+4)	1	1	0	0	1	0	0

	2^6 or 64	2^5 or 32	2^4 or 16	2^3 or 8	2^2 or 4	2^1 or 2	2^0 or 1
13 (=8+4+1)				1	1	0	1
29 (=16+8+4+1)			1	1	1	0	1
53 (=32+16+4+1)		1	1	0	1	0	1
127 (=64+32+16+8+4+2+1)	1	1	1	1	1	1	1

To a Measurable Infinity and Beyond

CALCULATING THE NUMBER OF ATOMS IN THE UNIVERSE

How many atoms are there in the universe? Believe it or not, regardless of how huge the figure must be, it's possible to express the number using just four digits.

$$\Im = 9^{9^{9^{9}}}$$

To write out that number in full would require $10^{369693100}$ digits, which, frankly, isn't possible when you consider the estimated age of the universe is less than 10^{18} seconds. $\Im$ came courtesy of German mathematician and physicist Carl Friedrich Gauss (1777–1855), who wanted a 'measurable infinity', which is a little odd given that in mathematics infinity is a concept rather than a number, denoting something that is beyond any fixed bound and which can't be resolved by counting or measurement, even in theory.

The symbol for infinity, ∞, was introduced by the English mathematician John Wallis in 1655, although discussion of the concept of infinity can be traced to as far back as the ancient Greeks.

Shaking Things Up

RICHTER SCALE: 6

Energy: 6.3 x 10^{13} J

Power of the atomic bomb dropped on Hiroshima.

THE ENERGY OF RICHTER

Earthquake magnitude is typically a measurement of ground motion, which is then expressed as a value on the Richter scale, developed in 1935 by American seismologist Charles F. Richter. It's based on powers of ten, meaning that an earthquake measuring 5 on the Richter scale has a recorded seismograph amplitude ten times greater than one measuring 4.

RICHTER SCALE: 6.1

Energy: 9 x 10^{13} J

Amount of energy in 1 gram of matter according to Einstein's celebrated $E = mc^2$.

Magnitude can also be translated into the seismic energy released by an earthquake, measured in joules (J). Here, an increase of 1 on the Richter scale represents a more than thirty-fold increase in the amount of seismic energy. This enables the power of earthquakes to be compared with other energy sources and vice versa.

As you can see, the most devastating earthquake ever recorded is not the earthquake with the greatest magnitude. There are plenty of other factors that have a bearing on how much damage an earthquake causes, such as the depth of the quake (a shallow ones is typically more devastating than a deep one), the physical environment, population density and the quality of building construction.

RICHTER SCALE: 6.3

Energy: 1.8 x 10^{14} J

Christchurch, New Zealand, earthquake, 2011.

RICHTER SCALE: 7

Energy: 2 x 10^{15} J

Haiti earthquake, 2010.

RICHTER SCALE: 7.5

Energy: 1 x 10^{16} J

Impact energy that formed meteor crater in Arizona.

RICHTER SCALE: 8

Energy: 6.3 x 10^{16} J

Approximate value of the 1556 Shaanxi earthquake, China – the most devastating earthquake in recorded history, estimated to have killed more than 830,000 people.

The Richter Scale

Amplified Seismic Ground Motion (Microns)

10^9
10^8
10^7
10^6
10^5
10^4
10^3
10^2
10^{-1}

Great

Major

Strong

Moderate

Small

Minor

Magnitude

0 1 2 3 4 5 6 7 8 9 10

A graphical representation of the Richter scale. Each level represents a change of 10 x the seismic amplitude and over 30 x the energy released.

Did we know the mechanical affections of the particles of rhubarb, hemlock, opium, and a man, as a watchmaker does those of a watch ... we should be able to tell beforehand that rhubarb will purge, hemlock kill, and opium make a man sleep.

JOHN LOCKE (1632–1704), ENGLISH PHILOSOPHER
AN ESSAY CONCERNING HUMAN UNDERSTANDING

RICHTER SCALE: 8.3
Energy: 1.7×10^{17} J
Total energy from the Sun that hits the Earth every second.

RICHTER SCALE: 9.6
Energy: 1.5×10^{19} J
Total consumption of electrical energy in the US in 2018.

RICHTER SCALE: 8.4
Energy: 2.4×10^{17} J
Tsar Bomba, the largest nuclear bomb ever detonated.

RICHTER SCALE: 9
Energy: 2×10^{18} J
Tohoku earthquake off the east coast of Japan, 2011.

RICHTER SCALE: 8.3
Energy: 1.7×10^{17} J
Estimated energy released by the eruption of Krakatoa in 1883.

And the Winner Is ...

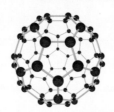

C_{60} (Molecule of the year, 1991)

RECOGNISING THE GREATEST ACHIEVEMENTS IN SCIENCE

In 1989, echoing *Time* magazine's Man of the Year (now known as Person of the Year), the journal *Science* began awarding the title of Molecule of the Year. The aim was to honour 'the scientific development of the year most likely to have a major impact on scientific advances and societal benefits'.

If we're being pedantic, the winner wasn't always a molecule but was sometimes a process. In 1995, there was a further shift when the award was given to a 'state of matter' that had been hypothesised seventy years earlier and finally produced that year. Perhaps not surprisingly, the award was subsequently given the more all-embracing title Breakthrough of the Year. Here's a list of the winners:

Molecule of the Year

1989	PCR (polymerase chain reaction), a technique that enables many copies of DNA to be generated from a tiny amount
1990	Manufacture of synthetic diamonds
1991	C_{60} (buckminsterfullerene), an allotrope of carbon that looks a bit like a football
1992	NO (nitric oxide) and its role in how cells in the body operate
1993	p53, a protein found in the body, which was recognised for its potential as a tumour suppressor
1994	The DNA repair enzyme system
1995	Bose–Einstein condensate (a state of matter)

Breakthrough of the Year

1996	Advances in the understanding of HIV disease
1997	The cloning of Dolly the sheep
1998	The ever-increasing expansion of the universe and the increasing evidence of dark matter
1999	The potential of stem cells
2000	The mapping of the human genome
2001	Nano, or molecular, circuits
2002	Small RNAs (ribonucleic acids) and their role in genome regulation
2003	The realisation that the universe is mostly made of dark energy
2004	The 'rover' missions to Mars
2005	Evolution in action
2006	Proving the Poincaré conjecture
2007	Human genetic diversity – looking at genomes individually
2008	Reprogramming cells through the insertion of genetic material
2009	4.4 million year-old skeleton is shown to be a human ancestor
2010	The first quantum computer
2011	Antiretroviral drugs reduce heterosexual transmission of HIV
2012	The discovery of the Higgs Boson
2013	Cancer immunotherapy
2014	Rosetta comet mission
2015	CRISPR genome-editing method
2016	The first observation of gravitational waves
2017	The first full observation of a neutron-star merger l
2018	Single cell sequencing
2019	The first image of a black hole is created
2020	Covid-19 vaccines developed at incredible speed

Geological Time Piece

4.275b yrs
22:47 hrs
First dinosaurs appear

4.35b yrs
22:56 hrs
First mammals appear

4.44b yrs
23:40 hrs
Last mass extinction, dinosaurs disappear

23:59:12 hrs
Genus *Homo* appears

23:59:56 hrs
Modern humans

4.05b yrs
21:36 hrs
First land plants appear

4.1b yrs
21:52 hrs
First insects appear

4b yrs
21:20 hrs
First fish appear

3,97b yrs
21:10 hrs
First trilobites appear

3.5b yrs
18:40 hrs
First plant life appears, probably in the form of green algae

4.25b yrs
22:40 hrs
Permian–Triassic extinction, the largest in Earth's history

30m yrs
00:10 hrs
Earth and planet Theia collide. Debris from the collision results in the Moon

2b yrs
10:41 hrs
Great Oxygenation Event when oxygen was no longer captured by oceans or land. So begins significant increase in the atmosphere.

1.5b yrs
08:00 hrs
Last common universal ancestor (LUCA)

1.7b yrs
9:04 hrs
Bacteria begin to produce oxygen

400–700m yrs
02:08–03:44 hrs
Late heavy bombardment

A NEW PERSPECTIVE ON EARTH'S EVOLUTION

Sometimes, discovering when something happened only begins to make sense when it's seen in relation to other events. This is particularly easy to do if the total time is shown in the form of a clock. The age of Earth is about 4.5 billion years – on the clock, one second equals approximately 52,000 years and one hour is 187.5 million years.

'Twin limb-like basalt columns'

ANTIQUE ATTRACTION
Sphinx of Ramesses II from
his Temple of Amun at Wadi
es-Sebua.

TWISTING SHELLEY'S WORDS

I met a traveller from an antique land,
Who said: 'Two vast and trunkless legs of stone
Stand in the desert. Near them on the sand,
Half sunk, a shattered visage lies, whose frown
And wrinkled lip and sneer of cold command
Tell that its sculptor well those passions read
Which yet survive, stamped on these lifeless things,
The hand that mocked them and the heart that fed.
And on the pedestal these words appear:
'My name is Ozymandias, King of Kings:
Look on my works, ye mighty, and despair!'
Nothing beside remains. Round the decay
Of that colossal wreck, boundless and bare,
The lone and level sands stretch far away.

'OZYMANDIAS'
PERCY BYSSHE SHELLEY (1792–1822)

You may very well have encountered Percy Bysshe Shelley's sonnet
'Ozymandias' somewhere before – it's a much-anthologised poem,
not to mention the inspiration for The Sisters of Mercy's cracking
song 'Dominion', which includes the final line of the poem in its
lyrics. Being so ubiquitous, it's perhaps not surprising to discover
that the poem has also come under scientific scrutiny.

The following spoof is an extract from a comic letter from
English geologist Neville Seymour Haile that first appeared in the
correspondence pages of the journal *Nature* in 1977. Of course,
any scientist should strive to be precise and meticulous in
their methods, but this suggested rewrite
of Shelley's poem to make it acceptable
for publication in a scientific journal
does, I hope, show that scientists
are also able to laugh at their
own peculiarities.

The scientific man is merely the minister of poetry. He is cutting down the Western Woods of Time; presently poetry will come there and make a city and gardens. This is always so. The man of affairs works for the behoof and use of poetry. Scientific facts have never reached their proper function until they merge into new poetic relations established between man and man, between man and God, or between man and Nature.

**SIDNEY LANIER (1842–1881),
AMERICAN MUSICIAN AND WRITER
'THE LEGEND OF ST. LEONOR',
MUSIC AND POETRY**

Suggested rewritten manuscript (summary)

Twin limb-like basalt columns ('trunkless legs') near Wadi Al-Fazar, and their relationships to plate tectonics

IBN BATTUTA[1] AND P. B. SHELLEY[2]

In a recent field trip to north Hadhramaut, the first author observed two stone leg-like columns 14.7m high by 1.8m in diameter (medium vast, ASTM grade scale for trunkless legs) rising from sandy desert 12.5km southwest of Wadi Al-Fazar (Grid 474 753). The rock is a tholeiitic basalt (table 1); 45 analyses by neutron activation technique show that it is much the same as any other tholeiitic basalt (table 2). A large boulder 6m southeast of the columns has been identified as of the 'shattered visage' type according to the classification of Pettijohn (1948, page 72). Granulometric analysis of the surrounding sand shows it to be a multimodal leptokurtic slightly positively skewed fine sand with a slight but persistent smell of camel dung. Four hundred and seventy-two scanning electron photomicrographs were taken of sand grains and 40 are reproduced here; it is obvious from a glance that the grains have been derived from pre-cambrian anorthosite and have undergone four major glaciations, two subductions, and a prolonged dry spell. One grain shows unique lozenge-shaped impact pits and heart-like etching patterns which prove that it spent some time in upstate New York. There is no particular reason to suppose that the columns do not mark the site of a former hotspot, mantle plume, triple junction, transform fault, or abduction zone (or perhaps all of these).

Keywords: plate tectonics, subduction, obduction, hotspots, mantle plume, triple junction, transform fault, trunkless leg, shattered visage

[1] *School of Earth & Planetary Sciences, University of the Fertile Crescent*
[2] *formerly of University College, Oxford*

'Oh Death, where is thy sting?'

MARKING 'MOSQUITO DAY'

In 1902, a year after the inception of the Nobel Prize, Ronald Ross (1857–1932), a doctor, became the first British recipient of the award. Five years earlier, while combining his research with medical service in India, Ross had shown the *Anopheles* mosquito to be a necessary factor in the transmission of malaria. It was a finding that would have an enormous impact on the fight against the disease.

For Ross – a man who had failed the Licentiate of the Society of Apothecaries examination as a medical student – the Nobel Prize in Physiology or Medicine 1902 and his subsequent knighthood in 1911 marked quite a change of fortune. Also an accomplished writer – his memoirs won the James Tait Black Memorial Prize in 1923 – Ross wrote the following poem, now inscribed on a monument to Ross at the SSKM Hospital, in Kolkata, India, describing his famous discovery on 20 August 1897, which he later called 'Mosquito Day', and celebrated annually thereafter.

This day relenting God
Hath placed within my hand
A wondrous thing; and God
Be praised; At his command,
Seeking His secret deeds
With tears and toiling breath,
I find thy cunning seeds,
O million-murdering Death.
I know this little thing
A myriad men will save.
O Death, where is thy sting?
Thy victory, O Grave?

> It's OK to sleep with a hypothesis, but you should never become married to one.
>
> **ANONYMOUS**

Triple Letter Score!

THE ETYMOLOGY AND SCRABBLE™ SCORE OF SOME COMMON SCIENTIFIC WORDS

Word	From ...	Meaning	Score
Acid	Latin, *acidus*	Sour	7
Alkali	Arabic, *al qalíy*	'The roasted ashes', the name originally given to soda ash	10
Anode	Greek, *ana* and *hodos*	Up path – positive electrode towards which current flows	6
Anthrax	Greek, *anthrax*	Charcoal – the disease causes black spots to appear on the skin	17
Argon	Greek, *argos*	Lazy, idle – argon was difficult to isolate because it's so chemically unreactive	6
Baryon	Greek, *barys*	Heavy – at the time of their discovery, baryons had the greatest mass of the subatomic particles	11
Cathode	Greek, *kata* and *hodos*	Down path – negative electrode away from which current flows	13
Cell	Latin, *cella*	Small room	6
Chlorine	Greek, *chloros*	Pale green	13
Eclipse	Greek, *ek* and *leipo*	Failure to appear or leave out	11
Electron	Greek, *elektron*	Amber – when rubbed, amber becomes charged with static electricity	10
Energy	Greek, *en* and *ergon*	To work in	10

Word	From ...	Meaning	Score
Gas	Greek, *khaos*	Chaos	4
Genetic	Greek, *genesis*	Coming into being	10
Gravity	Latin, *gravis*	Heavy	14
Hadron	Greek, *adros*	Thick, bulky	10
Helium	Greek, *helios*	The Sun – helium was first discovered through spectral analysis of a solar eclipse	11
Hydrogen	Greek, *hydro* and *gene*	Water producer	16
Malaria	Italian, *mala* and *aria*	Bad air	9
Mitochondria	Greek, *mitos* and *chondros*	Thread-like grains	20
Molecule	Latin, *moles* and *culus*	Small mass	11
Nucleus	Latin, *nucleus*	Kernel	9
Plankton	Greek, *planktos*	Wanderer	14
Proton	Greek, *protos*	The first	8
Quantum	Latin, *quantum*	How much	18
Science	Latin, *scientia*	Knowledge	11
Species	Latin, *species*	Appearance, kind	11

Taking Acid

UNDERSTANDING THE PH SCALE

Short for 'power of hydrogen', pH is a logarithmic scale regarding the concentration of hydrogen ions (H^+) in a solution – the lower the pH, the greater the concentration of hydrogen ions. The scale is used to measure how acidic or alkaline a solution is.

Pure water has a pH of 7.0 – which refers to a hydrogen ion concentration of 10^{-7} mol dm^{-3} (you don't need to worry exactly what that means) – and is considered neutral. Any solution with a pH less than 7 is said to be acidic and anything above, alkaline.

Because the scale is logarithmic, when the concentration of hydrogen ions changes by a factor of ten, the pH changes by only one unit. To put that in perspective, we can look at the powerhouses of Britain's 18th- and 19th-century industrial revolution, such as Manchester and Huddersfield. These two places are less than 25mi (40.2km) from the peat moorland at

Bleaklow in the Peak District, now an English national park. A century and a half of sulphurous emissions from these cities have resulted in the park's peat now having a pH of 2 (a hydrogen ion concentration of 10^{-2} mol dm^{-3}), as opposed to the typical peat value of 4 (or a hydrogen ion concentration of 10^{-4} mol dm^{-3}). The difference between 2 and 4 on the scale perhaps doesn't sound like much, but in fact it means the concentration of hydrogen ions in Bleaklow peat is 100 times greater than ordinarily expected levels.

The pH scale was originally developed by the Danish scientist Søren Sørenson (1868–1939) while working at the Carlsberg Laboratory in Copenhagen. Shown below are the approximate pH values of some everyday solutions, several of which might surprise you, especially if you consider your teeth's enamel is deprived of essential minerals at pH 5.5 and below.

COLOUR-CODED
The colours of universal indicator along with the pH values of some common foods and household items.

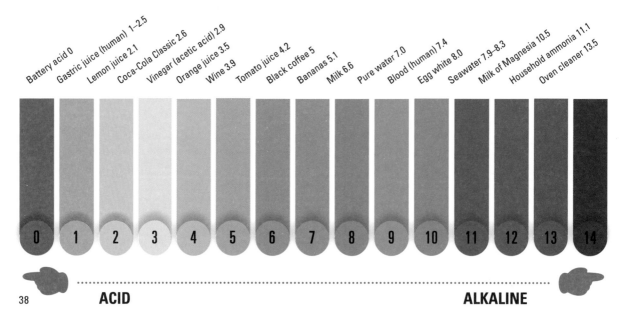

Battery acid 0 · Gastric juice (human) 1–2.5 · Lemon juice 2.1 · Coca-Cola Classic 2.6 · Vinegar (acetic acid) 2.9 · Orange juice 3.5 · Wine 3.9 · Tomato juice 4.2 · Black coffee 5 · Bananas 5.1 · Milk 6.6 · Pure water 7.0 · Blood (human) 7.4 · Egg white 8.0 · Seawater 7.9–8.3 · Milk of Magnesia 10.5 · Household ammonia 11.1 · Oven cleaner 13.5

0 1 2 3 4 5 6 7 8 9 10 11 12 13 14

ACID **ALKALINE**

A Good Indicator of …

A DIY PH TEST

Using just a red cabbage, a kettle, a pan and some glass jars or bottles, it's possible to make a pH indicator solution that will let you know whether something is acidic or alkaline, as well as indicating roughly to what degree. Here's how:

Slice and dice the red cabbage and put it in the pan. Now cover it with freshly boiled water and give it a good stir. Let it stand for a few minutes.

Use a sieve to remove the cabbage, leaving you with just the liquid. This is your indicator. It should be stored in a clean bottle, preferably in the dark.

When you want to use the indicator to test something, put some in a clear jar and then add whatever it is you're testing, such as apple juice, baking soda or even colourless shampoo. Make sure you always have a 'control'; that is, some indicator to which nothing has been added so that it's easy to see the change caused in other jars by whatever you're testing. Observe the change in colour that occurs when you add the substance you're testing to the indicator. The table below gives an approximate guide to the pH that the various possible colours correspond to:

Colour	Pink	Dark red/purple	Violet	Blue	Blue-Green	Green-Yellow
Approximate pH	1–2	2–3	5–7	8	9–10	11–12

The indicator works because red cabbage contains a pigment belonging to a class of molecules called anthocyanins. This pigment changes colour depending on the concentration of H^+ ions as they very slightly change its chemical structure. As such, the same species of cabbage can be planted in different locations and, depending on the pH of the different soils, produce cabbage crops that differ in colour.

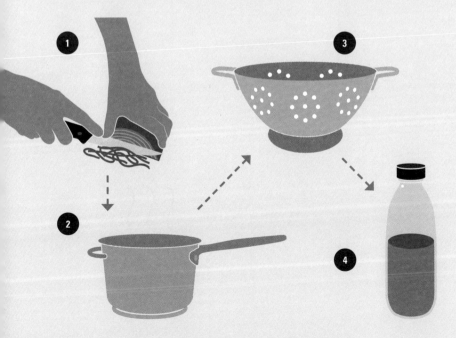

MAKING THE INDICATOR

1. Roughly chop the cabbage.
2. Steep in hot water for a few minutes.
3. Strain and cool.
4. Pour into a bottle with a lid, and use as the indicator.

Solar Eclipses Galore

WHERE AND WHEN TO SEE AN ECLIPSE

Throughout history, solar eclipses have captured the human imagination more than any other astronomical event. From Homer's *Odyssey* to H. Rider Haggard's *King Solomon's Mines*, they have often featured in literature as dire portents. That particular power has perhaps now passed, but witnessing an eclipse is still an incredible experience. The following gives an idea of where the next few eclipses will appear.

● Total Eclipse

● Annular* Eclipse

*An annular eclipse occurs when the Moon is too far from the Earth to obscure the Sun completely. As a result, a very bright ring surrounding the Moon is observed.

UPCOMING ECLIPSES
The paths of various eclipses that are due to take place up to the year 2040.

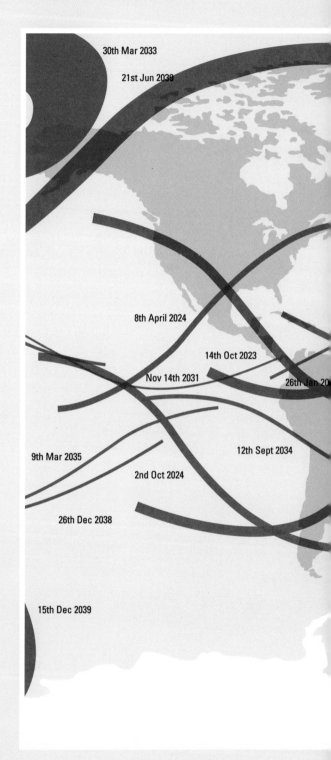

30th Mar 2033
21st Jun 2039
8th April 2024
14th Oct 2023
Nov 14th 2031
26th Jan 20
9th Mar 2035
12th Sept 2034
2nd Oct 2024
26th Dec 2038
15th Dec 2039

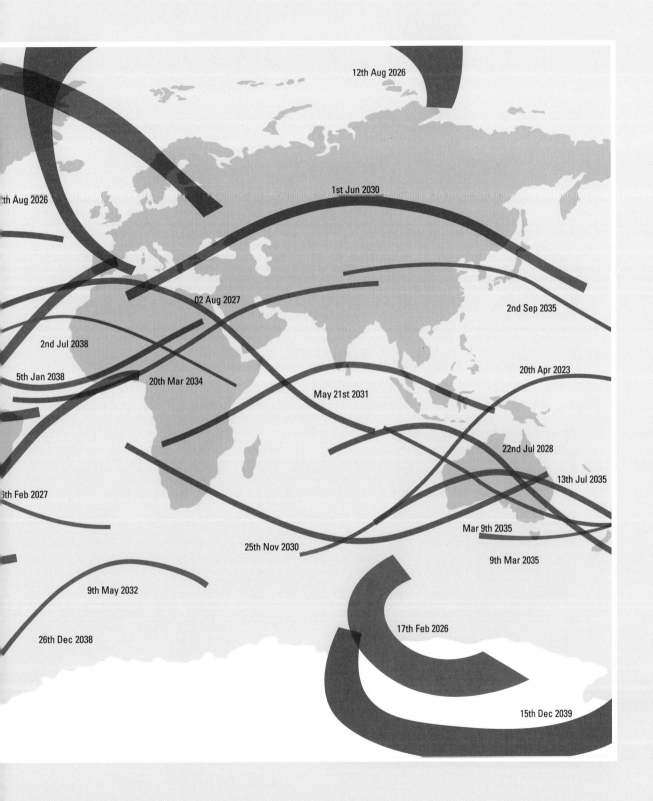

12th Aug 2026

th Aug 2026

1st Jun 2030

02 Aug 2027

2nd Sep 2035

2nd Jul 2038

5th Jan 2038

20th Mar 2034

20th Apr 2023

May 21st 2031

22nd Jul 2028

13th Jul 2035

6th Feb 2027

Mar 9th 2035

25th Nov 2030

9th Mar 2035

9th May 2032

17th Feb 2026

26th Dec 2038

15th Dec 2039

I Holy See the Error of My Ways

GALILEO'S BRUSH WITH THE CHURCH

Born in the Italian city of Pisa in 1564, Galileo Galilei is now widely regarded as one of the most important scientists ever to have lived. But in his own lifetime, he was persecuted by the Roman Catholic Church for daring to propound what is now a generally accepted truth about the structure of the solar system.

Galileo left university in 1585 without the degree in medicine for which he'd been studying. However, since he showed an exceptional aptitude for mathematics, his family were able to help him secure a professorship in this subject four years later. By 1610, he was living in Venice, and it was in this year that *The Sidereal Messenger* was published – a book describing the discoveries Galileo had made using his own improved version of the recently invented telescope. Included in the work – which quickly brought him to the attention of Europe's intelligentsia – were observations of the moons of Jupiter, the uneven surface of the Earth's moon and the existence of more stars in the heavens than could be perceived with the naked eye.

Galileo Galilei

The Church feted Galileo for his findings the following year. However, it was becoming clear that the Ptolemaic view of the solar system embraced by the Church – in which the Sun, the Moon and all the planets orbited the Earth – was no longer supported by the evidence. This left two alternatives: one that had been published by Polish astronomer Nicolaus Copernicus almost seventy years previously, and one put forward by the Danish astronomer Tycho Brahe in 1583.

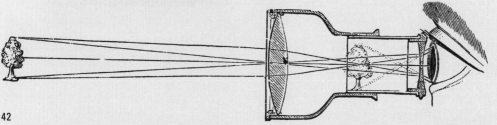

A Galilean telescope is defined as having one convex lens and one concave lens.

HELIOCENTRIC

GEOCENTRIC

DIFFERENT PATHS
Heliocentric paths revolve around
the Sun, whereas geocentric ones
have the Earth at their centre.

GEO- OR HELIOCENTRIC?

Let's look at Brahe's theory first. His astronomical observations led him to stick
with a stationary Earth at the centre of things, but with only the Sun and the Moon
orbiting it, while the other planets now orbited the Sun. As far as the Church was
concerned, this was OK – it was still consistent with a literal reading of the Bible
(references such as Ecclesiastes 1:5, 'The Sun also arises, and the Sun goes down, and
hastens to the place where it arose'), while also accounting for many of the anomalies
in Ptolemy's arrangement, which had dominated thought for the previous 1,400
years. By contrast, Copernicus's system had the Sun at the centre and the Earth
orbiting it, along with all the other planets.

Galileo and the German astronomer Johannes Kepler (more on him later in the
book) argued for the Copernican system, while pretty much everyone else supported
Brahe's theory. One of the key issues in the debate was to what extent the Bible was
the whole truth and nothing but the truth. Since his theory seemed to directly
contradict the Bible, in 1616 the Church condemned Copernicus, and any book
propounding his work was banned. Galileo was effectively given a warning. However,
he took the view that the Church needed to be convinced of its mistake.

DEFENDING HIS THESIS
Cristiano Banti's (1824–1904) painting *Galileo facing the Roman Inquisition.*

In August 1623, Cardinal Maffeo Barberini, a supporter of Galileo, became Pope Urban VIII. A few months later, Galileo dedicated his book *The Assayer* to the new pontiff, filling the scientist with the confidence to begin a new work, *Dialogue Concerning the Two Chief World Systems: Ptolemaic and Copernican*, intending to persuade non-specialists to his view of a heliocentric (sun-centred) system. The Church's censors, the Congregation of the Index, allowed the publication of the book in 1632, but when the Pope came to read it, he's reported to have 'exploded into great anger'. On 12 April 1633, Galileo was forced to stand trial. Incredibly, his defence was a denial of holding a Copernican view. On 22 June 1633, Galileo signed the recantation (reproduced on the next page) and was sentenced to house arrest. In failing health and going blind, he went on to write what many consider his masterpiece – *Discourses on Two New Sciences*. Galileo died on 8 January 1642.

The English poet and polemicist John Milton, author of *Paradise Lost*, visited Galileo during his house arrest. This event would later feature in Milton's *Areopagitica*, a stirring defence of free speech and attack on censorship. It's fair to say, the Church's actions towards Galileo would go down as one of the worst PR disasters in its history and a building block in the belief of some that there's an inherent antagonism between science and religion.

I, Galileo Galilei, son of the late Vincenzio Galilei of Florence, aged seventy years, being brought personally to judgment, and kneeling before you, Most Eminent and Most Reverend Lords Cardinals, General Inquisitors of the Universal Christian Republic against heretical depravity, having before my eyes the Holy Gospels which I touch with my own hands, swear that I have always believed, and, with the help of God, will in future believe, every article which the Holy Catholic and Apostolic Church of Rome holds, teaches and preaches. But because I have been enjoined, by this Holy Office, altogether to abandon the false opinion which maintains that the Sun is the centre and immovable, and forbidden to hold, defend, or teach, the said false doctrine in any manner [...] I am willing to remove from the minds of your Eminences, and of every Catholic Christian, this vehement suspicion rightly entertained towards me, therefore, with a sincere heart and unfeigned faith, I abjure, curse and detest the said errors and heresies, and generally every other error and sect contrary to the said Holy Church; and I swear that I will never more in future say, or assert anything, verbally or in writing, which may give rise to a similar suspicion of me; but that if I shall know any heretic, or anyone suspected of heresy, I will denounce him to this Holy Office, or to the Inquisitor and Ordinary of the place in which I may be. I swear, moreover, and promise that I will fulfil and observe fully all the penances which have been or shall be laid on me by this Holy Office. But if it shall happen that I violate any of my said promises, oaths and protestations (which God avert!), I subject myself to all the pains and punishments which have been decreed and promulgated by the sacred canons and other general and particular constitutions against delinquents of this description. So, may God help me, and His Holy Gospels, which I touch with my own hands, I, the above named Galileo Galilei, have abjured, sworn, promised and bound myself as above; and, in witness thereof, with my own hand have subscribed this present writing of my abjuration, which I have recited word for word.

Rabbit, Rabbit, Rabbit-rabbit, Rabbit-rabbit-rabbit, Rabbit-rabbit-rabbit-rabbit-rabbit

THE MAGIC OF SEQUENTIAL NUMBERS

In 1202, the mathematician Leonardo of Pisa, better known today as Fibonacci, published *Liber Abaci* (*The Book of Calculation*). It was via this book that Hindu numerals (the numeric symbols we still use today) were first introduced to Europe. It also contained a problem regarding the breeding of rabbits that was to have a considerable impact on the history of mathematics.

Fibonacci asked how many pairs of rabbits could be produced in a year from a single pair in an enclosed space if each pair produced a new breeding pair every month, including the initial pair in the first month; each new pair started reproducing at the age of one month; and no rabbit ever died.

The answer, it turned out, is 377 (see the table on the left).

Month	1	2	3	4	5	6	7	8	9	10	11	12
Adult pairs	1	2	3	5	8	23	21	34	55	89	144	233
New pairs	1	1	2	3	5	8	13	21	34	55	89	144
Total pairs	2	3	5	8	13	21	34	55	89	144	233	377

What's interesting about the above isn't the actual answer, but rather the sequence of numbers in the three rows – they're basically the same and they're what we now refer to as the Fibonacci sequence:

1, 1, 2, 3, 5, 8, 13, 21, 34, 55, 89, 144, 233, 377…

And something incredibly fascinating lies within this sequence.

If you look at the ratio of successive pairs of Fibonacci numbers, placing the larger number first (for example, 144/89 and 233/144), as we get further along the sequence the ratio between the two gets closer and closer to 1.618, or, more precisely, ½ (1+√5). This is commonly known as the golden ratio, Φ.

46

The golden ratio is often cited as an aesthetically pleasing proportion and has appeared extensively in the arts during the past two and a half thousand years. The French architect Le Corbusier used the ratio in some of his 20th-century buildings, the Spanish surrealist artist Salvador Dali did the same in his 1955 painting *The Sacrament of the Last Supper*, and for a time book page proportions were based on it (but not this one, sadly). There are also many other occasions where the ratio's appearance happens more by chance and indeed is only approximate.

Much more startling than its use by humans, however, is the frequent occurrence of the Fibonacci sequence and golden ratio in nature. The number of petals on flowers is often a Fibonacci number. If you look at the head of a sunflower, the florets of a cauliflower, the bumps on a pineapple or the scales of a pine cone, you'll notice a particular spiral pattern – so many in one direction and a different number in the other. Nearly always, these two numbers will be consecutive Fibonacci numbers, which we earlier saw are the best whole number approximations of Φ. And when it comes to the leaves of a plant, Φ crops up in the typical angle (137.5°) between adjacent leaves. Known as the 'golden angle', its connection with Φ is: 360°/$\Phi \approx$ 222.5°, which is the same as 360° – 137.5°. This arrangement, which in three dimensions is spiral, is believed to give the best possible exposure to light for every leaf, taking into account their position in relation to each other. Wow!

NATURALLY OCCURRING
A golden spiral, which gets further from its starting point by a factor of φ for every quarter turn it makes.

Ban Dihydrogen Monoxide!

FAKE SCIENTIFIC NEWS

The following scientific spoof by Craig Jackson is his version of an idea first originated by three fellow students at the University of California, Santa Cruz, in 1989. It's a superb example of how scientific language and terminology, along with judicious selection and disingenuous presentation of 'facts', can be used to confuse the unprepared. Everything below is, strictly speaking, true. But once you realise exactly what molecule is being discussed, any fear simply evaporates!

THE INVISIBLE KILLER

Dihydrogen monoxide is colourless, odorless, tasteless, and kills uncounted thousands of people every year. Most of these deaths are caused by accidental inhalation of DHMO, but the dangers of dihydrogen monoxide do not end there. Prolonged exposure to its solid form causes severe tissue damage. Symptoms of DHMO ingestion can include excessive sweating and urination, and possibly a bloated feeling, nausea, vomiting, and body electrolyte imbalance. For those who have become dependent, DHMO withdrawal means certain death.

DIHYDROGEN MONOXIDE:

- is also known as hydroxyl acid, and is the major component of acid rain
- contributes to the 'greenhouse effect'
- may cause severe burns
- contributes to the erosion of our natural landscape
- accelerates corrosion and rusting of many metals
- may cause electrical failures and decreased effectiveness of automobile brakes
- has been found in excised tumors of terminal cancer patients.

CONTAMINATION IS REACHING EPIDEMIC PROPORTIONS!

Quantities of dihydrogen monoxide have been found in almost every stream, lake, and reservoir in America today. But the pollution is global, and the contaminant has even been found in Antarctic ice. DHMO has caused millions of dollars of property damage in the Midwest, and recently California.

DESPITE THE DANGER, DIHYDROGEN MONOXIDE IS OFTEN USED:

- as an industrial solvent and coolant
- in nuclear power plants
- in the production of styrofoam
- as a fire retardant
- in many forms of cruel animal research
- in the distribution of pesticides – even after washing, produce remains contaminated by this chemical
- as an additive in certain 'junk-foods' and other food products.

Companies dump waste DHMO into rivers and the ocean, and nothing can be done to stop them because this practice is still legal. The impact on wildlife is extreme, and we cannot afford to ignore it any longer!

IT'S TOO LATE!

Act NOW to prevent further contamination. Find out more about this dangerous chemical. What you don't know can hurt you and others throughout the world.

THE HORROR MUST BE STOPPED!

The American government has refused to ban the production, distribution or use of this damaging chemical due to its 'importance to the economic health of this nation'. In fact, the navy and other military organizations are conducting experiments with DHMO, and designing multi-billion-dollar devices to control and utilize it during warfare situations. Hundreds of military research facilities receive tons of it through a highly sophisticated underground distribution network. Many store large quantities for later use.

DIHYDROGEN MONOXIDE
=
WATER (H_2O)

'A remarkable book, sure to make a mighty stir'

DARWIN'S MASTERPIECE

Charles Darwin's *On the Origin of Species* was published on 24 November 1859. Despite costing almost as much as a policeman's weekly wage, the book rapidly sold out the near 1,200 copies available for sale from its first print run.

The initial critical reception of *On the Origin of Species*, a tome of almost mythic status in the history of science, makes interesting reading. Out of the daily newspapers, only *The Times*, *The Morning Post* and *The Daily News* reviewed it, along with a similar number of weeklies. The *News of the World* contained a notice with an extract from the book on slave-making ants! But the review journals, which played a vital role in public debate, covered the book in significant force. What follows are the opening two paragraphs of a review that appeared in *The Examiner* on 3 December 1859, less than two weeks after the book was published. One can't help but admire the reviewer's evenhandedness, particularly as three weeks later biologist Thomas Henry Huxley, later known as 'Darwin's bulldog' because of his vociferous support of Darwin's theory, wrote in his review in *The Times* that it was currently impossible to 'affirm absolutely either the truth or falsehood of Mr Darwin's views', and that it could take another twenty years to determine whether Darwin was right or had, in fact, 'over-estimate[d] the value of his principle of natural selection' (privately, though, Huxley is said to have remarked 'how stupid of me not to have thought of that' after he first read Darwin's book).

On the Origin of Species by Means of Natural Selection, or the Preservation of Favoured Races in the Struggle for Life.

By Charles Darwin, M.A., F.R.S., Author of 'Journal of Researches During a Voyage Round the World.' John Murray, Albermarle Street, 1859.

'This is a remarkable book, sure to make a mighty stir among the philosophers – perhaps even among the theologians. Indeed the very reputation of such a work from such an authority as Mr Darwin would seem to have done so already, for, if we are rightly informed, the entire edition was taken off on the first day of publication. Those who have perused his 'Voyage round the World' need not be told that the author is a man of curious and careful research, familiar with every branch of natural knowledge and gifted with the faculty of expressing himself, even on questions often abstruse, in language always perspicuous and often eloquent.

'Mr Darwin's work, although extending to 500 pages, is but an abstract of a greater which he is preparing, and which two or three years hence will be completed.

'The doctrine he adopts to account for the present condition of the living world is, in fact, a revival of an old one of the transmutation of species; but he illustrates it with an amount of knowledge and ingenious appliances never before brought to its support. We are ourselves by no means convinced by his reasoning, nor do we think that it overthrows the existing theory of philosophers, founded on the evidences of geological discoveries, that the organic world, as we see it, is the result of a succession of creations and destructions. There will, however, no doubt, be many converts to Mr Darwin's opinions, which for the perfect integrity with which they are stated are entitled to the most respectful study.'

BACKLASH

'A mighty stir' it most definitely did create. Vociferous opponents included the palaeontologist and soon to be founder of the Natural History Museum in London Richard Owen. He passionately believed in the idea of 'fixed species', which had been the cornerstone of his scientific work. But this was something that the *Origin* by definition rejected. Owen's work was a major influence on the then Bishop of Oxford, Samuel Wilberforce, whose infamous 'debate' with Huxley at the 1860 meeting of the British Association for the Advancement of Science contained a quip from the bishop about man being descended from apes, which was intended to ridicule. Huxley recalled the celebrated episode, along with his reply: 'If then, said I, the question is put to me would I rather have a miserable ape for a grandfather or a man highly endowed by nature and possessed of great means of influence & yet who employs these faculties & that influence for the mere purpose of introducing ridicule into a grave scientific discussion, I unhesitatingly affirm my preference for the ape.'

Even more remarkable, among the answers to questions at a recent examination for the General Certificate of Education at Advanced level was the following: 'Darwin's theory was based upon three good solid pints [sic]; 1. the struggle for exits 2. the survival of the fattest 3. maternal election'.

FROM SIR GAVIN RYLANDS DE BEER'S
CHARLES DARWIN: EVOLUTION BY
***NATURAL SELECTION* (1963)**

The Truth Will Out ... Eventually

T.H. Huxley

'THE FOUR STAGES OF PUBLIC OPINION'

Darwin's defender T. H. Huxley was a firm believer that *magna est veritas et praevalebit* ('truth is great and will prevail') but, as he drolly remarked, 'truth is great, certainly, but considering her greatness, it is curious what a long time she is apt to take about prevailing'. He expanded upon this sentiment in a notebook jotting regarding 'The Four Stages of Public Opinion'. It's hard to disagree with his analysis.

I
JUST AFTER PUBLICATION

The Novelty is absurd and subversive of Religion and Morality.

The Propounder both fool and knave.

II
TWENTY YEARS LATER

The Novelty is absolute Truth and will yield a full and satisfactory explanation of things in general.

The Propounder man of sublime genius and perfect virtue.

III
FORTY YEARS LATER

The Novelty won't explain things in general after all and therefore is a wretched failure.

The Propounder a very ordinary person advertised by a clique.

IV
A CENTURY LATER

The Novelty is a mixture of truth and error. Explains as much as could reasonably be expected.

The Propounder worthy of all honour, in spite of his share of human frailties, as one who has added to the permanent positions of science.

The Swiss-American palaeontologist Louis Agassiz (1807–1873) made a similar observation when he said, 'Every great scientific truth goes through three stages. First, people say it conflicts with the Bible. Next, they say it had been discovered before. Lastly, they say they always believed it'.

'Endless forms most beautiful'

DARWIN'S GRAND VIEW OF LIFE

The final paragraph of *On the Origin of Species* is worthy of an entry of its own, since it sums up Darwin's thesis so beautifully, and succinctly points out that a scientific understanding of the world need not denude it of its wonder – it can, in fact, enhance it.

It is interesting to contemplate an entangled bank, clothed with many plants of many kinds, with birds singing on the bushes, with various insects flitting about, and with worms crawling through the damp earth, and to reflect that these elaborately constructed forms, so different from each other, and dependent on each other in so complex a manner, have all been produced by laws acting around us. These laws, taken in the largest sense, being Growth with Reproduction; Inheritance, which is almost implied by Reproduction; Variability from the indirect and direct action of the external conditions of life, and from use and disuse; a Ratio of Increase so high as to lead to a Struggle for Life, and as a consequence to Natural Selection, entailing Divergence of Character and the Extinction of less-improved forms. Thus, from the war of nature, from famine and death, the most exalted object which we are capable of conceiving, namely, the production of the higher animals, directly follows. There is grandeur in this view of life, with its several powers, having been originally breathed into a few forms or into one; and that, whilst this planet has gone cycling on according to the fixed law of gravity, from so simple a beginning endless forms most beautiful and most wonderful have been, and are being, evolved.

Are You Cleverer than a Fifteen-year-old … from 1858?

TAKING A VICTORIAN TEST

A national curriculum was first introduced into the UK in 1988. Before that, a school's choice of examination board determined what a student was to learn for that subject. The first of these boards started in 1858, when both the universities of Oxford and Cambridge ran local public examinations for children aged under sixteen (junior candidates) and under eighteen (senior candidates) for the first time.

The Cambridge exams were held in eight locations, including Birmingham, Grantham, Liverpool and Norwich. However, these exams were only public and egalitarian up to a point – the cost to sit one was £1, which, like the cover price of *On the Origin of Species*, was an amount close to a policeman's weekly wage. Set by the university's dons, the junior examinations included papers on 'pure mathematics', 'mechanics and hydrostatics', 'chemistry' (both theoretical and practical) and 'zoology and botany'. In total, there were ten sections examined, and every student had to pass three while entering no more than six. Religious knowledge, the first of the sections, was compulsory, unless the student's parents or guardians objected. Ironically, the examiner's first report mentioned there being 'some difficulty in settling the examination in Religious Knowledge owing to the great difference of opinion which appeared to prevail concerning it'.

Many of the mathematics questions make sense to us today, even if a few are beyond the expected knowledge of contemporary sixteen-year-olds, especially when they involve log tables. This is true also of some of the science questions, although they also serve to illustrate how much was yet to be discovered and how scientific language has changed.

On the opposite page are some sample questions from the junior paper.

PURE MATHEMATICS

- Explain how to find the sum of a series of n terms in arithmetical progression, whose first term is a, and last term l.
- Sum the series: 6, −2, ⅔, −²⁄₉, &c. to infinity.

....

MECHANICS AND HYDROSTATICS

- Find the ratio of the power to the weight in a system of pullies where all the strings are attached to a uniform bar from which the weight is suspended, the weights of the pullies being neglected. From what point of the bar ought the weight to be suspended that the bar may rest in a horizontal position?
- A is a fixed pully, B, C heavy moveable pullies. An inextensible string without weight is thrown over A. One end of it passes under C and is fastened to the centre of B, the other end passes under B and is fastened to the centre of A. Compare the weights of B and C that the system may be in equilibrium, the strings being all parallel.

....

CHEMISTRY

- Name the different compounds of Nitrogen and Oxygen, and state the composition of each. By what qualities could you recognise the deutoxide (or binoxide) of Nitrogen?

....

ZOOLOGY

- Which Vertebrates are oviparous; which are abranchiate; and which have gills during only a period of their existence? Which mammals have the simplest kind of teeth?

....

BOTANY

- What parts of plants are the: Blackberry, Strawberry, Mulberry, Apple, Potato, Beet, Tea and Opium?

Just a Pinch of Vanadium

MAGNESIUM
0.027%

HYDROGEN
10%

CARBON
23%

NITROGEN
2.6%

COBALT
0.000003%

SODIUM
0.14%

OXYGEN
61%

WHAT ARE YOU MADE OF?

This is the composition of the essential
elements in the average adult human body
by mass (before any multivitamin and mineral
tablets have been swallowed):

CALCIUM
1.4%

URANIUM
0.0000001%

POTASSIUM
0.2%

PHOSPHORUS
1.1%

MOLYBDENUM
0.00001%

COPPER
0.00010%

SILICON
0.026%

TIN
0.00002%

VANADIUM
0.00001%

FLUORINE
0.0037%

CHROMIUM
0.000003%

IRON
0.006%

IODINE
0.00002%

MANGANESE
0.00002%

NICKEL
0.00001%

ZINC
0.0033%

SULPHUR
0.2%

Taxonomically Speaking

BIOLOGICAL CLASSIFICATION

Organisms are classified according to their similarities and, by extension, differences. By definition, they satisfy the first classification, life, which is then followed by their placement within a domain, of which there are three: Archaea, Bacteria and Eukarya. Then there are five kingdoms – animals, bacteria, fungi, plants and protists (for example, amoeba and plasmodium, which causes malaria) protoctista (for example, sponges and seaweed). The taxonomic hierarchy then continues down through the following categories: phylum, class, order, family (sometimes with the subgroup tribe), genus and finally species, of which millions exist. The table below shows how the hierarchy works with humans, lions, pacific bluefin tuna and the damask rose as examples.

	Humans	Lion	Pacific bluefin tuna	Damask rose
Kingdom	Animalia	Kingdom	Animalia	Plantae
Phylum	Chordata	Chordata	Chordata	Embryophyta
Class	Mammalia	Mammalia	Actinopterygii	Magnoliopsida
Order	Primates	Carnivora	Perciformes	Rosales
Family	Hominidae	Felidae	Scombridae	Rosaceae
(Tribe)	Hominini		Thunnini	
Genus	Homo	Panthera	Thunnus	Rosa
Species	H. sapiens	P. leo	T. orientalis	R. damascena

Biodiversity Under Pressure

THREATENED SPECIES

The International Union for Conservation of Nature (IUCN) was founded in 1948 and regularly produces the Red List of Threatened Species, which gives information on the 'status of wild species and their links to livelihoods'. Species are categorised using a system of nine labels:

EX – **Extinct**
EW – **Extinct in the Wild**
CR – **Critically Endangered**
EN – **Endangered**
VU – **Vulnerable**
NT – **Near Threatened**
LC – **Least Concern**
DD – **Data Deficient**
NE – **Not Evaluated**

A threatened species is defined as any listed as Critically Endangered (CR), Endangered (EN) or Vulnerable (VU). The following table gives you some idea of the scope of the survey and shows the number of species currently classed as threatened.

In July 2020, it was reported that more than 96 per cent of the 107 species of lemur should be classed as threatened, with 33 of them Critically Endangered – ten more than in 2012.

	Estimated number of described species	Number of species evaluated by 2021	Number of threatened species in 2021
Vertebrates			
Mammals	6,578	5,968	1,333
Birds	11,162	11,162	1,445
Reptiles	11,690	10,148	1,839
Amphibians	8,395	7,296	2,400
Fishes	36,058	22,581	3,332
SUBTOTAL	**73,883**	**67,166**	**10,437**
Invertebrates			
Insects	1,053,578	12,100	2,270
Molluscs	83,706	9,019	2,385
Crustaceans	80,122	3,189	743
Corals	5,610	818	232
Arachnids	110,615	441	251
Velvet worms	208	11	9
Horseshoe crabs	4	4	2
Others	157,543	902	150
SUBTOTAL	**1,491,386**	**26,514**	**6,042**
TOTAL	**1,565,269**	**93,680**	**16,479**

The Big Five

MASS EXTINCTION EVENTS

A mass extinction is defined as the disappearance of a dramatic number of families (typically more than 10 per cent) or species (usually more than 40 per cent) in a relatively short space of time (in relation to the geological timescale, so in actual fact we could still be talking about hundreds of thousands of years). It's believed that there have been more than twenty such events in Earth's history, but in 1982, American palaeontologists Jack Sepkoski and David M. Raup proposed a list of 'The Big Five'.

In total, 99 per cent of species that have ever lived on Earth no longer exist.

Extinction	When	Estimated size/impact	Main theory regarding cause
Ordovician–Silurian mass extinction	Approx. 440 million years ago	86% of species	Climate cooling due to glaciation
Late Devonian mass extinction	Approx. 360 million years ago	75% of species	Changes in sea level, a decrease in oxygen concentrations in the sea, global cooling
Permian mass extinction	Approx. 250 million years ago	96% of species	Volcanism, global warming and ocean acidification
Triassic–Jurassic mass extinction	Approx. 200 million years ago	80% of species	Increase in atmospheric CO_2 leading to global warming and ocean chemistry change
Cretaceous–Tertiary mass extinction	Approx. 65 million years ago	76% of species	Extraterrestrial impact (asteroid/meteor) leading to rapid cooling

Evolutionary Patterns

THE LAST 500 MILLION-YEAR EVOLUTION OF VERTEBRATES

In his book *Vertebrate Palaeontology*, Professor Mike Benton of the University of Bristol includes a spindle diagram showing 'the pattern of evolution of the vertebrates'. With his kind permission it's reproduced here.

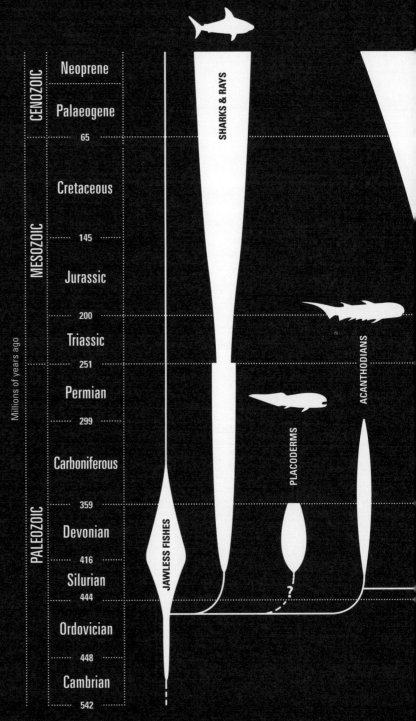

Millions of years ago

CENOZOIC	Neoprene	
	Palaeogene	
	65	
MESOZOIC	Cretaceous	
	145	
	Jurassic	
	200	
	Triassic	
	251	
	Permian	
	299	
PALEOZOIC	Carboniferous	
	359	
	Devonian	
	416	
	Silurian	
	444	
	Ordovician	
	448	
	Cambrian	
	542	

SHARKS & RAYS

ACANTHODIANS

PLACODERMS

JAWLESS FISHES

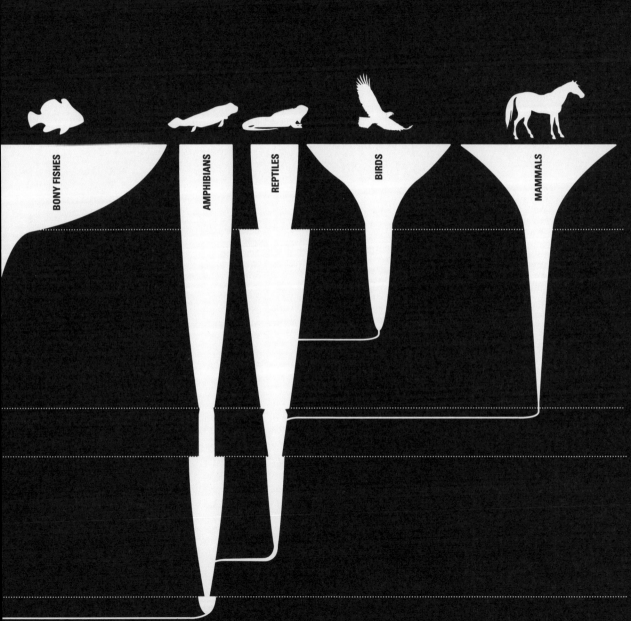

BONY FISHES

AMPHIBIANS

REPTILES

BIRDS

MAMMALS

The Torino Impact Hazard Scale

NO HAZARD

White Zone / Torino Scale: 0

The likelihood of a collision is zero or is so low as to be effectively zero. Also applies to small objects such as meteors and bodies that burn up in the atmosphere, as well as infrequent meteorite falls that rarely cause damage.

PLOTTING THE PROBABILITY OF AN EXTRATERRESTRIAL COLLISION

In the 1979 disaster movie *Meteor*, an asteroid 8km (5 miles) wide is on a collision course with the Earth. The planet's only hope is a collaboration between the USA and the Soviet Union, involving their respective orbiting nuclear missile launchers. It's fair to say that the film isn't the finest hour for its all-star cast, but it does serve to highlight the potential for disaster from outer space (the Cretaceous-Tertiary – K-T – event that wiped out the dinosaurs and much more perhaps serves as a more highbrow example).

In 1979, there was no formalised way of expressing the seriousness of such a situation. It would be another sixteen years before, in 1995, Professor Richard P. Binzel first presented the then unnamed Torino Impact Hazard Scale to provide an effective way of assessing the potential effect upon the Earth of any newly discovered asteroid or comet and then communicating that to the public. Expect to find it cropping up in some disaster movie or other sooner or later. The scale works by plotting the kinetic energy of the extraterrestrial object (half its mass × its expected velocity squared) against the probability of it striking the Earth. This value can change over time as information about the object improves.

NORMAL

Green Zone / Torino Scale: 1

A routine discovery in which a pass near the Earth is predicted that poses no unusual level of danger. Current calculations show the chance of collision is extremely unlikely with no cause for public attention or concern. New telescopic observations very likely will lead to reassignment to level 0.

MERITING ATTENTION BY ASTRONOMERS

Yellow Zone / Torino Scale: 2

A discovery, which may become routine with expanded searches, of an object making a somewhat close but not highly unusual pass near the Earth. While meriting attention, there's no cause for public concern as a collision is very unlikely. New telescopic observations very likely will lead to reassignment to level 0.

MERITING ATTENTION BY ASTRONOMERS

Yellow Zone / Torino Scale: 3

Current calculations give a 1 per cent or greater chance of collision capable of localised destruction. Most likely, new telescopic observations will lead to reassignment to level 0. Attention by public and by public officials is merited if the encounter is less than a decade away.

MERITING ATTENTION BY ASTRONOMERS

Yellow Zone / Torino Scale: 4

Current calculations give a 1 per cent or greater chance of collision capable of regional devastation. Most likely, new telescopic observations will lead to reassignment to level 0. Attention by public and by public officials is merited if the encounter is less than a decade away.

Copyright ©1999, 2004 Richard P. Binzel, Massachusetts Institute of Technology.

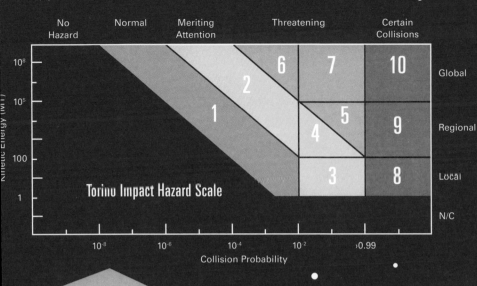

Torino Impact Hazard Scale

The scale works by plotting the kinetic energy of the extraterrestrial object (mv^2 i.e. mass x its expected velocity squared) against the probability of it striking the Earth. This value isn't necessarily fixed; it can change over time as information regarding the object improves. At the time of writing, there were two objects with a Torino Scale value of 1 and nothing higher, so there's no need to panic.

THREATENING
Orange Zone / Torino Scale: 5

A close encounter posing a serious but still uncertain threat of regional devastation. Critical attention by astronomers is needed to determine conclusively whether or not a collision will occur. If the encounter is less than a decade away, governmental contingency planning may be warranted.

THREATENING
Orange Zone / Torino Scale: 6

A close encounter by a large object posing a serious but still uncertain threat of a global catastrophe. Critical attention by astronomers is needed to determine conclusively whether or not a collision will occur. If the encounter is less than three decades away, governmental contingency planning may be warranted.

THREATENING
Orange Zone / Torino Scale: 7

A very close encounter by a large object, which, if occurring this century, poses an unprecedented but still uncertain threat of a global catastrophe. For such a threat in this century, international contingency planning is warranted, especially to determine urgently and conclusively whether or not a collision will occur.

CERTAIN COLLISION
Red Zone / Torino Scale: 10

A collision is certain, capable of causing global climatic catastrophe that may threaten the future of civilisation as we know it, whether impacting land or ocean. Such events occur on average once per 100,000 years, or less often.

CERTAIN COLLISION
Red Zone / Torino Scale: 8

A collision is certain, capable of causing localised destruction for an impact over land or possibly a tsunami if close offshore. Such events occur on average between once per fifty years and once per several thousand years.

CERTAIN COLLISION
Red Zone / Torino Scale: 9

A collision is certain, capable of causing unprecedented regional devastation for a land impact or the threat of a major tsunami for an ocean impact. Such events occur on average between once per 10,000 years and once per 100,000 years.

At the time of writing, there were no objects with a Torino scale value higher than 0, so there's no need to panic.

It Has Been Long Known that …

LIFTING THE LID ON SCIENTIFIC JARGON

Have you ever wondered what certain phrases in scientific papers really mean? Well, in 1957, C. D. Graham, who worked in a General Electric Company research laboratory, lifted the lid when he wrote the following 'glossary for research reports' in the journal *Metal Progress*.

INTRODUCTION

It has been long known that …	I haven't bothered to look up the original reference
… of great theoretical and practical importance	… interesting to me
While it has not been possible to provide definite answers to these questions …	The experiments didn't work out, but I figured I could at least get a publication out of it

EXPERIMENTAL PROCEDURE

The W-Pb system was chosen as especially suitable to show the predicted behaviour …	The fellow in the next lab had some already made up
High purity … Very high purity … Extremely high purity … Superpurity … Spectrascopically pure …	Composition unknown except for the exaggerated claims of the supplier
A fiducial reference line …	A scratch
Three of the samples were chosen for detailed study …	The results on the others didn't make sense and were ignored
… accidentally strained during mounting	… dropped on the floor
… handled with extreme care throughout the experiments	… not dropped on the floor

> The most exciting phrase to hear in science, the one that heralds new discoveries, is not, 'Eureka!' ('I found it!') but rather, 'Hmm … that's funny …'
>
> **ISAAC ASIMOV (1920–1992), AMERICAN SCIENCE FICTION WRITER AND BIOCHEMIST**

Wrong Imaginary Non-existent
I think Fair

RESULTS

Typical results are shown …	The best results are shown
Although some detail has been lost in reproduction, it is clear from the original micrograph that …	It is impossible to tell from the micrograph
Presumably longer times …	I didn't take time to find out
The agreement with the predicted curve is excellent	Fair
Good	Poor
Satisfactory	Doubtful
Fair	Imaginary
… as good as could be expected	Non-existent
These results will be reported at a later date	I might possibly get around to this sometime
The most reliable values are those of Jones	He was a student of mine

DISCUSSION

It is suggested that … It is believed that … It may be that …	I think
It is generally believed that …	A couple of other guys think so too
It might be argued that …	I have such a good answer to this objection that I shall raise it now
It is clear that much additional work will be required before a complete understanding …	I don't understand it
Unfortunately, a quantitative theory to account for these effects has not been formulated	Neither does anyone else
Correct within an order of magnitude	Wrong
It is hoped that this work will stimulate further work in the field	This paper isn't very good, but neither are any of the others in this miserable subject

ACKNOWLEDGEMENTS

Thanks are due to Joe Glotz for assistance with the experiments and to John Doe for valuable discussions	Glotz did the work and Doe explained what it meant

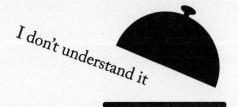

I don't understand it

The $11 Million Book

AUDUBON'S *THE BIRDS OF AMERICA*

At 5pm on 7 December 2010, the London-based auctioneer Sotheby's began a session consisting of ninety-one lots entitled 'Magnificent Books, Manuscripts and Drawings from the Collection of Frederick 2nd Lord Hesketh'. These included a first folio edition of Shakespeare's *Comedies, Histories & Tragedies* from 1623, and Ranulph Higden's *Polychronicon*, printed by William Caxton in 1482. But the undoubted star of the collection was a first edition, comprising four volumes, of *The Birds of America*, which had been published by the author, John James Audubon, between 1827 and 1838. It sold for more than $11 million and became the most expensive book ever bought at auction (the previous holder of the record was also *The Birds of America* and, reporting on the sale, *The Economist* newspaper revealed that 'a list of the ten most expensive books would include five copies of *The Birds of America*').

There are only 120 copies of *The Birds of America* still known to exist. Measuring roughly 97cm × 65cm (38.2in x 25.6in), the book is a collection of 435 magnificent hand-coloured etched plates, with line engraving and aquatint, detailing many of the bird species of the United States, including some now extinct. Its production costs were so high that it was published in eighty-seven sets of five plates over an eleven-year period. It is, quite simply, a work of art. It's also an ornithological masterpiece and had a profound influence on many natural historians, including Charles Darwin, who would go on to refer to Audubon three times in his *On the Origin of Species*.

Opposite is a list of the seven most expensive science books bought at auction.

Title	Author	Year edition published	Value
The Birds of America	John James Audubon (1785–1851)	1827–1838	$10.3 million (2010)
Les Liliacées	Pierre-Joseph Redouté (1759–1840)	1802	$5 million (1985)
Geographia – Cosmographia	Claudius Ptolemy (c. AD 90–c. AD 168)	1462 edition	$3.5 million (2006)
De revolutionibus orbium coelestium	Nicolaus Copernicus (1473–1543)	1543	$1.9 million (2008)
Complete Folio of Birds	John Gould (1804–1881)	1831–1888	$1.8 million (1998)
De humani corporis fabrica	Andreas Vesalius (1514–1564)	1543	$1.5 million (1998)
Principia Mathematica	Isaac Newton (1642–1727)	1687	$3.7 million (2016)

It's a Laughing Matter

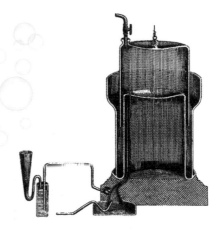

THE POETS' SCIENTIST

Aged not quite twenty and already showing significant scientific promise, Cornishman Humphry Davy joined the Pneumatic Institution in Bristol in 1798. During the following year, he investigated the effects of nitrous oxide, or laughing gas, which English chemist Joseph Priestley had discovered more than twenty years earlier. The Scottish inventor James Watt built Davy a portable gas chamber and was one of the sixty 'breathers' who took part in the studies, along with Davy's good friends the poets Samuel Taylor Coleridge and Robert Southey. As well as recording and publishing the results of his experiments, Davy penned the following lines to describe the effects of the gas, which he seems to have rather enjoyed:

> *Not in the ideal dreams of wild desire*
> *Have I beheld a rapture-wakening form:*
> *My bosom burns with no unhallow'd fire,*
> *Yet is my cheek with rosy blushes warm;*
> *Yet are my eyes with sparkling lustre fill'd;*
> *Yet is my mouth replete with murmuring sound;*
> *Yet are my limbs with inward transports fill'd,*
> *And clad with new-born mightiness around.*

Poetry was to prove a lifelong love for Davy. His poems were published in an anthology in 1799, including one he wrote aged just seventeen, and Davy was asked by poet William Wordsworth to proofread the second edition of his *Lyrical Ballads* (1800). Coleridge would later write that Davy was 'the Man who *born* first a Poet first converted Poetry into Science'.

As a scientist, Davy built on the work of French chemist Antoine Lavoisier, whose book *Traité élémentaire de chimie* (*Elementary Treatise on Chemistry*), published in 1789, was considered the first true chemistry book. As well as inventing, most famously, a lamp that allowed miners to work much more safely by dramatically reducing the risk of underground explosions, Davy developed the method of electrolysis, using this to isolate elements such as potassium and calcium for the first time. He also gave chlorine its name and was an acclaimed populariser of science, lecturing to sell-out crowds. Davy succeeded the botanist Joseph Banks as president of the Royal Society in 1820 and is reported to have humbly said that his greatest discovery was his assistant Michael Faraday, who would go on to eclipse him in the annals of science. Ultimately, the lines of the last stanza of 'The Sons of Genius', the anthologised poem he wrote when seventeen, proved accurate of both himself and Faraday:

> *Theirs is the glory of a lasting name,*
> *The meed of Genius and her living fire;*
> *Theirs is the laurel of eternal fame,*
> *And theirs the sweetness of the muse's lyre.*

When Left Can Be Right and Right Can Be Wrong

THE CHIRALITY OF SMELL

And now for something about mirrors and molecules. It's a curious fact that the molecules chiefly responsible for the respective smells of spearmint and caraway seeds are exactly the same, except for one crucial difference.

The smell inducers are mirror images of the same molecule, carvone. This feature in chemistry is known as chirality, from the Greek word *kheir*, meaning 'hand'. Your hands are chiral. Your left hand is the mirror image of your right, but you can't superimpose one upon the other. So, in relation to carvone, we might say that the right-hand version is responsible for the smell of caraway seeds and the left-hand version is responsible for that of spearmint. This rather implies that our smell receptors are chiral too, otherwise

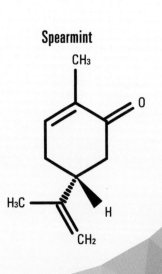

Spearmint

we wouldn't be able to distinguish between the two molecules, and caraway and spearmint would therefore smell the same. Think of each molecule as a Yale key and the other as its mirror image. Both keys won't fit the same lock.

It was French chemist Louis Pasteur who first noticed this feature when, using just a pair of tweezers, he painstakingly separated two different forms of tartaric acid crystals – left-hand and right-hand versions – and discovered they each rotated polarised light to the same degree but in opposite directions. A fifty–fifty mixture produced no effect at all, the two cancelling each other out.

Chirality crops up in all sorts of areas of nature, including drug synthesis. When a drug such as ibuprofen is made, it typically comprises a mixture of left- and right-hand versions, called a racemic mixture, which can have important consequences. It's highly likely that only one of these two forms will do the job it's intended to do. The ideal scenario would be that the other version has no effect whatsoever and is eventually excreted by the body. But at worst, it can have adverse side effects. This is exactly what happened with the drug thalidomide in the late 1950s. It was administered as an anti-nausea drug to pregnant women in its racemic form. Disastrously, the left-hand version of the molecule caused defects and deformities in foetuses growing in the womb. Drug companies now pay a significant amount of attention to the effects different chiral versions of the drug they manufacture produce, spending vast amounts of money on producing versions that include the drug in only its pure right-hand or left-hand form.

Caraway

Energetic Thoughts

ENERGY STORES AND TRANSFERS

At school, I was taught about different types, or forms, of energy – electrical and kinetic, for example. However, as the 20th-century American physicist Richard Feynman pointed out, we don't know what energy is and must instead think of it as 'a numerical quantity that does not change when something happens'.

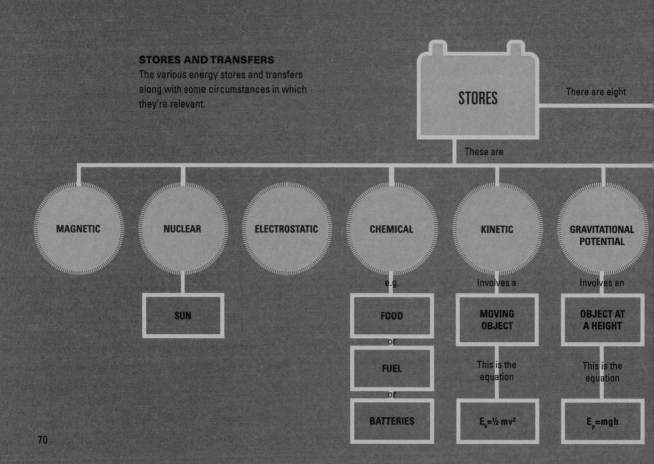

STORES AND TRANSFERS
The various energy stores and transfers along with some circumstances in which they're relevant.

STORES

There are eight

These are

MAGNETIC	NUCLEAR	ELECTROSTATIC	CHEMICAL	KINETIC	GRAVITATIONAL POTENTIAL
	SUN		e.g. FOOD or FUEL or BATTERIES	Involves a MOVING OBJECT — This is the equation $E_k = \frac{1}{2}mv^2$	Involves an OBJECT AT A HEIGHT — This is the equation $E_p = mgh$

At the turn of this century, Robin Millar, a professor in the Department of Education at the University of York, began to argue that education should move away from describing forms or types of energy. Instead, he proposed referring to energy stores and transfers. Using money as an analogy for energy, if I had £100, I might have it stored in a bank, under my mattress or in a piggy bank. I could transfer some or all of it electronically, by cheque or even by hand.

Students now learn about eight energy stores and four transfers, the names of which are essentially the same as those names used in the past for types or forms of energy. The stores are chemical, kinetic, gravitational potential, elastic potential, internal (thermal), nuclear, magnetic and electrostatic. The transfers are electrically, radiation (for example, light and sound), work done (whenever something is moved a distance) and heating. All stores and transfers have the unit Joule (J).

Stores and transfers make it easier to understand the various ways energy can be distributed. For example, energy in the nuclear store of the Sun is transferred by radiation to the Earth, where the process of photosynthesis stores it chemically.

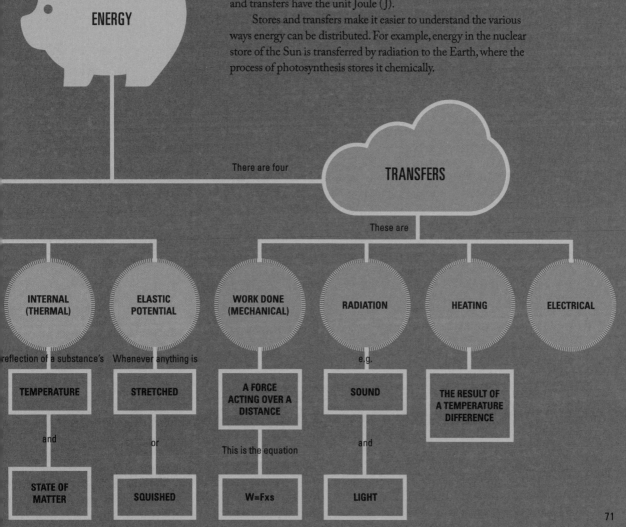

ENERGY

There are four

TRANSFERS

These are

| INTERNAL (THERMAL) | ELASTIC POTENTIAL | WORK DONE (MECHANICAL) | RADIATION | HEATING | ELECTRICAL |

reflection of a substance's

Whenever anything is

e.g.

| TEMPERATURE | STRETCHED | A FORCE ACTING OVER A DISTANCE | SOUND | THE RESULT OF A TEMPERATURE DIFFERENCE |

and

or

This is the equation

and

| STATE OF MATTER | SQUISHED | W=Fxs | LIGHT |

A Fully Rounded Member of Society

THE SCIENTIFIC EQUIVALENT OF: HAVE YOU READ A WORK OF SHAKESPEARE'S?

The English novelist C. P. Snow, a former research scientist and civil servant, gave the annual Rede Lecture at the University of Cambridge in 1959. Entitled 'The Two Cultures and the Scientific Revolution', it not only gave birth to a book, *The Two Cultures*, that has never been out of print since, it also launched a public debate that has perhaps never been satisfactorily settled – how the arts (or humanities) and science can best work with and support each other, particularly with regard to directing public and government policy.

At one point in *The Two Cultures*, Snow describes occasions when he has been in the company of 'highly educated' people with a scathing attitude towards scientists who have 'never read a major work of English literature'. Occasionally, Snow felt the need to respond by asking how many of them could describe the second law of thermodynamics? According to Snow, their subsequent incredulity showed that they failed to see his point that this was the equal of their disparagement – it was, he felt, 'the scientific equivalent of: have you read a work of Shakespeare's?'.

Snow was not merely sniping back at the detractors of science; his overall point was, to borrow a phrase from Shakespeare, that these two cultures, and society at large, would benefit most by being 'two distincts, division none'.

The next entry in this book will, among other things, describe the second law of thermodynamics. Since there wasn't room to include a full work by the Bard in these pages, you'll have to take the initiative and read one yourself – if you haven't already. After that, with luck you'll be a fully rounded individual, ready to take your place as a valued member of society.

You Can't Win, You Know

THE FOUR LAWS OF THERMODYNAMICS

In his *Autobiographical Notes*, Albert Einstein argued that 'a theory is the more impressive the greater the simplicity of its premises is, the more different kinds of things it relates, and the more extended is its area of applicability'. The example he gave of a theory demonstrating these qualities was classical thermodynamics, which Einstein was 'convinced … would never be overthrown'.

The Industrial Revolution and creation of the heat engine kick-started the concerted study of thermodynamics – a word not coined until Scottish-Irish engineer and physicist William Thomson, later Lord Kelvin, put it forward in an 1843 paper. Within ten years of this, versions of what would become the first and second laws of thermodynamics had been proposed, both framed within a classical view of thermodynamics, meaning they were based on what was empirically measurable, typically within a laboratory setting.

The 1870s and beyond saw the development of statistical thermodynamics due to work done by scientists such as Scottish physicist James Clerk Maxwell and Austrian physicist Ludwig Boltzmann. Here, probability theory was employed to understand the laws at the microscopic level of atoms and molecules. Around the same time, chemical thermodynamics, relating to chemical reactions and states of matter, was born. These two branches of thermodynamics would eventually lead to the proposal of the third law in 1912 by German chemist Walther Nernst.

The zeroth law was the fourth and last to be formally recognised, although the primacy of its name is testament to the fact it had been stated in various guises before and is something fundamental to the other three laws.

Each of the four laws can be stated in a variety of ways, typically depending on the branch of thermodynamics and the context in which it's being described, such as classical, statistical or chemical. The versions that follow are essentially the form in which the laws were first stated:

If someone points out to you that your pet theory of the universe is in disagreement with Maxwell's equations – then so much the worse for Maxwell's equations. If it is found to be contradicted by observation – well, these experimentalists do bungle things sometimes. But if your theory is found to be against the second law of thermodynamics, I can give you no hope; there is nothing for it but to collapse in deepest humiliation.

ARTHUR EDDINGTON (1882–1944),
ENGLISH ASTRONOMER, PHYSICIST AND MATHEMATICIAN
THE NATURE OF THE PHYSICAL WORLD (1928)

ZEROTH LAW

If two systems are in thermal equilibrium with a third system, they are also in thermal equilibrium with each other.

This means that all three systems would be at the same temperature. It's the zeroth law that allows a thermometer to do its job – it falls into thermal equilibrium with the system it's measuring, and we can read the temperature. If a third system is the same temperature, then all three are in thermal equilibrium with each other. The zeroth law also says that the net heat flow between two objects at the same temperature is zero.

FIRST LAW

Energy can be neither created nor destroyed, only interconverted between forms.

In effect, the first law is based on the principle of conservation of energy. For example, if you walk up a snowy mountain carrying a pair of skis, some of the chemical energy stored in your body is converted into potential energy (and almost certainly heat – make sure you wear layers), which is, in essence, a store of energy. If you now put your skis on and journey back down, this potential energy is converted into kinetic energy (this is energy related to motion). Other forms of energy include electrical, light, sound and, as Albert Einstein showed with his famous equation $E = mc^2$, mass.

SECOND LAW

Heat energy always flows spontaneously from a hot to a cold system and never naturally the other way round.

If you place ice in warm water, heat flows into it from the surrounding water. As the word 'naturally' in the above definition implies, heat can flow from cold to hot, but energy is needed to make it happen. If you think about it, this is exactly how ice is made in a freezer – the required energy is supplied through electricity (electrical energy). One implication of the second law is that it's impossible to make a perfect heat engine in which all the heat generated, such as in a combustion engine, can be converted into work: some of the heat must be lost to any surrounding colder system.

THIRD LAW

A system can't be reduced to absolute zero (0 kelvin) in a finite number of steps.

This is probably the hardest of the laws to grasp and the least relevant to everyday life. Which probably explains why it wasn't chosen by C. P. Snow to bolster his argument.

A Hot and Cold Problem

THE MPEMBA EFFECT

First revealed in the *Physics Education* journal in 1969, this is a story to lift the heart, and is a powerful lesson to us all. In 1963, Erasto B. Mpemba was in his third year at secondary school in Tanzania. During a physics lesson, he asked his teacher about something that had been puzzling him – why was it that the ice-cream mixture he'd made and had been boiling had frozen quicker than his friend's more tepid mixture, even though both had been placed in the freezer at the same time? The boy was simply told he must have been confused.

Mpemba didn't just accept this brush-off, and the problem continued to trouble him. His conviction in the truth of his observations increased as friends who regularly made ice cream confirmed it was quicker to do so the hotter the mixture was when put in the freezer.

After passing his O level qualification a few years later, Mpemba found himself in high school. The first topic he studied there was heat. He asked his new teacher the 'hotter ice cream freezing faster' question, and again he was told he was confused and even guilty of believing in 'Mpemba physics' as opposed to real physics. The teacher didn't let this go and it became a running joke with both him and Mpemba's classmates. Despite this, Mpemba persisted and experimented further with water in beakers. His findings proved the same as before.

A VISITING PROFESSOR

One day, Dr D. G. Osborne of University College Dar es Salaam visited the school and, at the end of the session, invited questions from the students. Mpemba bravely asked, 'If you take two similar containers with equal volumes of water, one at 35°C and the other at 100°C, and put them into a refrigerator, the one that started at 100°C freezes first. Why?' The rest of the audience sniggered, but Dr Osborne asked Mpemba to confirm he'd performed the experiment before promising to try it himself, even though he thought him mistaken. Dr Osborne recognised the 'need to encourage students to develop questioning and critical attitudes' and felt that there could be a 'danger… [in] authoritarian physics'. Much to his surprise, Dr Osborne got the same results, as did the university students he put to work on the problem. Furthermore, no answer could be found in any scientific literature to Mpemba's observation.

Finally, in the paper in *Physics Education*, jointly authored by Mpemba and Dr Osborne, a request was made to readers for further information. It turned out that the question had already puzzled many acclaimed minds, including the philosophers Aristotle, Francis Bacon and René Descartes. Today, the problem is fittingly known as the Mpemba Effect. Towards the end of June 2012, the *Royal Society of Chemistry* began a competition to find the 'best and most creative explanation' for the effect. The winning entry listed four factors: evaporation, dissolved gases, the mixing of the water through convective currents and supercooling.

Something terrible then happened in 2016 – a paper was published in the journal *Nature* that declared there was 'no evidence to support meaningful observations of the Mpemba Effect'. Curiously, this didn't prevent further research into the effect and, in 2020, Professor William Zimmerman at the University of Sheffield explained the strange phenomena as being due to microbubbles. He also explained 2016's findings as being the result of the researcher in question using pure water, while Mpemba and Osborne had used tap water. Whether this proves to be the final word remains to be seen.

Throwing a
Tumblerful of Water
into the Sea

MAXWELL'S DEMON

On 11 December 1867, James Clerk Maxwell wrote a letter to
his close friend, the Scottish physicist and mathematician Peter
Guthrie Tait. In it he described a thought experiment that
violated the second law of thermodynamics. In Maxwell's eyes,
it was designed to show that the law was guaranteed only in a
statistical sense. To illustrate his experiment, Maxwell invented
a 'being', which, in 1874, William Thompson (later Lord Kelvin)
described as 'Maxwell's intelligent demon'.

Maxwell introduced the experiment in his book *Theory of Heat* in 1871 in this way:
    'One of the best established facts in thermodynamics is that it is impossible in a
system enclosed in an envelope which permits neither change of volume nor passage
of heat, and in which both the temperature and the pressure are everywhere the same,
to produce an inequality of temperature or pressure without the expenditure of work.
This is the second law of thermodynamics, and it is undoubtedly true as long as we
can deal with bodies only in mass and have no power of perceiving or handling the
separate molecules of which they are made up.'
    But Maxwell then asks us to imagine 'a being whose faculties are so sharpened
that he can follow every molecule in its course'. This being, 'whose attributes are still
essentially finite as our own, would be able to do what is at present impossible for us'.
This is the 'demon'. Maxwell continued:
    'We have seen that the molecules in a vessel full of air at uniform temperature
are moving with velocities by no means uniform, though the mean velocity of any
great number of them, arbitrarily selected, is almost exactly uniform.
    'Now let us suppose that such a vessel is divided into two portions, A and B,
by a division in which there is a small hole, and that a being, who can see the
individual molecules, opens and closes this hole, so as to allow only the swifter
molecules to pass from A to B, and only the slower ones to pass from B to A.
He will thus, without expenditure of work, raise the temperature of B and lower
that of A in contradiction to the second law of thermodynamics.'

MAXWELL'S DEMON IN ACTION
The demon chooses when to open the 'trap door' between the two sides, so that we move from thermal equilibrium to a colder side on the left and a hotter one on the right.

A   B

A   B

A TUMBLERFUL OF WATER

Writing again about his idea in a letter to English scientist John William Strutt, Maxwell summarised the significance of his idea as follows: 'The second law of thermodynamics has the same degree of truth as the statement that if you throw a tumblerful of water into the sea, you cannot get the same tumblerful of water out again.' That's to say, although in practical, real-life terms it's true, it's not *absolutely* true. In Maxwell's scenario, the demon is using only information to produce a temperature difference – and therefore the necessary ingredients to run a heat engine (a steam engine, for example) – from a system in thermal equilibrium. In 1929, the Hungarian-American physicist Leó Szilárd (who would four years later conceive the idea of the nuclear chain reaction) argued that the demon's processing of information on each molecule's temperature *had* to involve energy, more than was subsequently gained, meaning that the second law wasn't in fact violated. The implication, though, was that information and energy were related.

Rather astonishingly, in the journal *Nature Physics* in 2010, a team of Japanese scientists reported making a particle 'climb up a spiral staircase-like potential'; that is, it gained energy, through the use of a real-time feedback control, thereby gaining more energy than had been put in. While there is still a long way to go, their research points to the incredible idea of an 'information-to-heat engine'.

On the Other Side of Silence

MEASURING LOUDNESS

In July 2009, the heavy metal band Kiss reportedly achieved 136 decibels during a live performance in Canada, thereby making them one of the loudest bands in the world. As we will see in a moment, that should be beyond the threshold of pain, although many would argue Kiss already surpass that the moment they step on stage.

We measure loudness in decibels. A decibel is a unit used to quantify the ratio between two power levels, one of which is treated as the reference level. In the case of measuring loudness, this reference level is taken to be the threshold of human hearing. The (rather complicated-looking) formula for its calculation is:

$$\text{Number of decibels} = 10 \log_{10} \left(\frac{P}{P_0} \right)$$

where P is the level effectively being measured and P_0 is the reference. As such, a difference in loudness of 10 decibels, say from 30dB to 40dB, means that one sound is ten times louder than the other. A change of 1 decibel represents an increase in loudness of approximately 26 per cent. This gives a context to the figures you often hear about.

Decibels are also used as units in optics and electronics, as well as other disciplines, because they are so useful. But when you hear them mentioned, it will almost certainly be in relation to how loud something is.

dB	Sensation
0	Threshold of hearing
10	Breathing
20	Rustling leaves
30	Watch ticking 1m away
40	Birdsong
50	Quiet conversation
70	Loud conversation
80	Door slamming
110	Pneumatic drill
130	Threshold of pain

EUCLID'S ALGORITHM

Al-Khwarizmi was a 9th-century Arab mathematician and the originator of algebra (from the Arabic *al-jabr*, meaning 'the reunion of broken parts'). If you can recall long, painful hours spent struggling to balance equations in school, you might not feel particularly well disposed towards him, but he really did have a brilliant mind. He's reported to have once declared that 'with my two algorithms, one can solve all problems – without error, if God wills it!'

Algorithms are very, very useful. A series of algorithms can be used to solve the Rubik's Cube (you'll find lots of examples on the internet), the sales rankings on Amazon are established using algorithms and they're also used an awful lot in the world of finance. Needless to say, algorithms can be extremely complex. As such, their increasing use and influence has happened in tandem with the development of the modern computer.

The principle of an algorithm can be easily demonstrated using what's known as Euclid's algorithm (named after the ancient Greek mathematician, Euclid), which enables us to find the greatest common factor of two numbers by using a particular sequence of divisions. A factor is a number that can divide a specified number exactly – for example, 2 is a factor of 6 because it divides into it exactly three times. A common factor is a number that can divide two different numbers exactly, so 4 is a common factor of 12 and 28.

First, we need two numbers, say 192 and 42. We then divide the larger number (192) by the smaller (42) in the following way to give a multiple of the smaller number plus a remainder:

$$192 = (4 \times 42) + 24$$

We then divide 42 by the remainder, 24:

$$42 = (1 \times 24) + 18$$

Then 24 by the remainder, 18:

$$24 = (1 \times 18) + 6$$
$$18 = (3 \times 6) + 0$$

And so, this tells us that the greatest common factor of 192 and 42 is 6. Pretty nifty, eh?

Quite an Illuminating Lecture

'THE CHEMICAL HISTORY OF THE CANDLE'

Born the son of a journeyman blacksmith on the Surrey borders of London in 1791, Michael Faraday would go on to become one of the most celebrated scientists of the 19th century, particularly because of his work on electricity and magnetism, including the discovery that electricity could be produced by repeatedly moving a magnet through a coil of wire. More than fifty years after Faraday's death, the novelist Aldous Huxley felt compelled to say, 'Even if I could be Shakespeare, I think I should still choose to be Faraday.' That's quite a compliment.

Faraday came to scientist Humphry Davy's attention when, aged just 21 and having just completed his bookbinding apprenticeship, he presented Davy with a bound copy of the lecture notes he'd taken down when attending the great man's demonstrations at the Royal Institution. It was a brave move from a young man who had previously declared it 'impossible … to follow [Davy]. I should merely injure and destroy the beautiful and sublime observations that fell from his lips'. Davy apparently felt otherwise, recommending Faraday for a position as a laboratory assistant.

Faraday proved a quick learner. Within less than a year, from having felt he could never hope to be a speaker of Davy's quality, he was writing to his friend Benjamin Abbott, confidently describing what the necessary qualities of a good lecturer were. In 1825, Faraday would originate the Royal Institution's now-celebrated series of Christmas Lectures, going on to present nineteen series of them himself. One of these series was called 'The Chemical History of the Candle' and was eventually published as a book in 1861, proving hugely successful. It's still available to buy today (I recommend it). Over the course of six lectures, Faraday took the audience through the remarkable science behind an everyday item everyone knew. What follows is just a brief taste of its magic and something you may want to try at home.

There is [a] condition which you must learn as regards the candle, without which you would not be able fully to understand the philosophy of it, and that is the vaporous condition of the fuel [the wax]. In order that you may understand that, let me show you a very pretty, but very commonplace experiment. If you blow a candle out cleverly, you will see the vapour rise from it. You have, I know, often smelt the vapour of a blown-out candle – and a very bad smell it is; but if you blow it out cleverly, you will be able to see pretty well the vapour into which this solid matter is transformed. I will blow out one of these candles in such a way as not to disturb the air around it, by the continuing action of my breath; and now, if I hold a lighted taper two or three inches from the wick, you will observe a train of fire going through the air till it reaches the candle. I am obliged to be quick and ready, because, if I allow the vapour time to cool, it becomes condensed into a liquid or solid, or the stream of combustible matter gets disturbed.

When It Comes to What's in Your Genes, Size Isn't Everything

GENETIC COMPLICATION

An organism's genome refers to all the genes – sections of DNA involved in determining a particular physical or biological trait. These are contained in a single set of its chromosomes – pairs of rod-like structures found inside every developed cell in the organism's body, and which carry its DNA.

From the inception of life, the codified information in genes determines the way an organism grows and what it becomes. This being the case, one would tend to assume that the more complicated an organism is, the more genes it must have. But when genomes started to be mapped, scientists made the startling discovery that this isn't the case. Humans, for example, were originally estimated to have about 100,000 genes, but the true figure, which is still being refined, appears to be less than a quarter of this. By comparison, the genome of a particular species of rice contains more than twice as many genes as a human. The following information taken from a 2008 paper in the journal *Nature Education* shows how the number of genes varies surprisingly between different organisms:

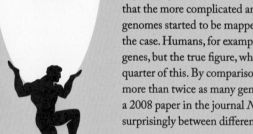

The scientist is not a person who gives the right answers, he is one who asks the right questions.

CLAUDE LÉVI-STRAUSS (1908–2009), FRENCH ANTHROPOLOGIST

Organism	Number of genes
Trichomonas vaginalis (a single-celled parasitic organism)	60,000
Oryza sativa (a common rice that feeds over half the world's population)	51,000
Mus musculus (mouse of choice for pet lovers and laboratory leaders)	30,000
Homo sapiens (humans)	20,000–25,000
Drosophila melanogaster (the common fruit fly, a staple of school experiments)	14,000
Saccharomyces cerevisiae (a yeast commonly used in baking and brewing)	6,000
Escherichia coli (commonly referred to as *E. coli*, this bacteria can cause food poisoning)	4,500

Would You Adam and Eve it?

MITOCHONDRIAL DNA

You've got some alien DNA in your body. Not alien in the sense of UFOs, but stuff that wasn't originally a part of you (or rather of your ancestors, way back). Mitochondria are found in pretty much every complex cell in the body. Consisting of tiny structures, called organelles, they have their own DNA (called, funnily enough, mitochondrial DNA). It's a legacy of the fact that mitochondria were once independent bacteria that evolved to live inside larger cells such as our own. So, although they are now an integral part of the way our bodies work – producing adenosine triphosphate (ATP), the currency of energy in all living organisms – technically, once upon a time they were just hitchhikers.

As if that weren't fascinating enough, another feature of mitochondrial DNA is that it's only inherited through the female line. This makes it possible for scientists to trace our maternal lineage, and such analysis has pointed to the incredible fact that we *all* share a single common ancestor who lived nearly 200,000 years ago. That's one female to whom every single one of us is related! Unsurprisingly, she's often referred to as 'mitochondrial Eve' or 'African Eve'.

Amazingly, there's also a male equivalent of mitochondrial Eve – 'Y-chromosome Adam'. This is because the Y chromosome is exclusive to males and is passed from father to son. As such, it has been possible for researchers to discover that all men are descended from a single male, again from Africa. As with mEve (as her friends call her), exactly where and when Y-Adam lived hasn't yet been fixed, although current studies suggest Y-Adam was some 60,000 years younger than mEve. The original toy boy then.

A man walks into a pub and asks for a pint of adenosine triphosphate.

The barman says, 'That's 80p!'

Particularly Taxing

MAKING THE STANDARD MODEL

OK, you might have to take a deep breath for this one –
I certainly did. Do you remember being taught that atoms
are made up of three subatomic particles: protons, neutrons
and electrons? The nucleus of an atom, found at its centre,
is made up of the neutrons and protons, and only these
contribute to an element's atomic mass. Both neutrons and
protons are roughly the same mass. The electrons, on the other
hand, 'orbit' the nucleus and have negligible mass in comparison
to it. But this description only scratches the surface.

Particle physics can be really said to have started in 1897 when English physicist
J. J. Thomson discovered the electron. As this area progressed over the following
years, it was often the case that theory preceded discovery. Wolfgang Pauli,
an Austrian physicist, proposed the existence of the neutrino in 1930, but it wasn't
until 1956 that it was formally detected. By the early 1960s, there was a veritable
'particle zoo', and it was believed that there existed a plethora of elementary particles.

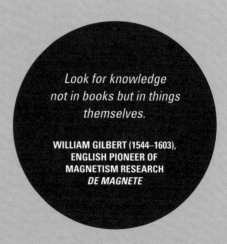

*Look for knowledge
not in books but in things
themselves.*

**WILLIAM GILBERT (1544–1603),
ENGLISH PIONEER OF
MAGNETISM RESEARCH
*DE MAGNETE***

In 1964, the American physicists Murray Gell-Mann and George Zweig proposed, independently of each other, a classification scheme, which became known as the quark model. Murray Gell-Mann had assigned the name 'quark' the previous year, taking inspiration for its spelling from a dream the chief protagonist has in the novel *Finnegans Wake*, by the Irish writer James Joyce, in which a drunken seagull says, 'three quarks for Muster Mark'. A more sober seagull would've asked for three quarts.

AN EMERGING STANDARD

In the early 1970s, scientists came together to develop what has become known as the Standard Model of particles and forces. It proposes that everything in the universe is made from twelve fundamental particles and four fundamental forces (technically, one of these forces, gravity, isn't in the standard model, as we'll soon see).

The model consists of three groups: quarks, leptons and gauge bosons, as well as the Higgs boson (we'll come to bosons in a bit). Quarks and leptons are what make up matter and are classed together as fermions. Within their respective groups, quarks and leptons are put in pairs, which are also known as generations. The first generation makes up stable matter.

Quarks come in six 'flavours': up (u), down (d), charm (c), strange (s), top (t) and bottom (b). Corresponding with each quark is an antiquark: $\overline{u}$, $\overline{d}$, $\overline{c}$, $\overline{s}$, $\overline{t}$ and $\overline{b}$.

When quarks are grouped in pairs, mesons are formed; when grouped in triplets, they form baryons. Protons and neutrons, which form the nucleus of an atom, are types of baryons. In turn, both baryons and mesons belong to the hadron family, which also includes antibaryons.

To put it another way, hadrons are all composed of different combinations of quarks, which are classed as either mesons, baryons or antibaryons. Baryons consist of three quarks, antibaryons three antiquarks and mesons a quark and an antiquark.

Protons are baryons that consist of u, u and d quarks. Neutrons are also a type of baryon and comprise u, d, and d quarks. High-energy collisions can create other hadrons beside protons and neutrons. This happens, for example, in the upper atmosphere, where cosmic rays (high-energy protons from outer space) collide with nuclei of oxygen and nitrogen to create a particle shower that includes mesons.

Electrons are examples of leptons, of which there are six flavours: electron (e^-), electron neutrino (v_e), muon (u^-), muon neutrino (v_u), tauon (τ^-) and tauon neutrino (v_τ). The electron, muon and tauon all have a charge of -1 and differ only in mass (electrons are smaller than muons, which are, in turn, smaller than tauons). Neutrinos have no charge.

For each lepton, there's a corresponding antilepton, which has a charge opposite to that of the matching lepton.

So, there you have it, a somewhat exhausting whistle-stop tour of the particle zoo, including the twelve elementary

u Up	c Charm	t Top	g Gluon
d Down	s Strange	b Bottom	γ Photon
e Electron	μ Muon	τ Tau	Z Z Boson
Ve Electron neutrino	Vμ Muon neutrino	Vτ Tau neutrino	W W Boson
			H Higgs Boson

BUILDING BLOCKS
The fundamental particles of the Standard Model – the building blocks of all matter.

- Quarks
- Leptons
- Gauge bosons
- Scalar boson

particles from which all matter in the universe is composed. Now, let's take a look at the four fundamental forces of nature while you still have the will to live.

THE FOUR FUNDAMENTAL INTERACTIONS OF NATURE AND GAUGE BOSONS

The strong force – This is the force between quarks that allows the formation of protons and neutrons. It also acts between protons and neutrons themselves and is responsible for giving the nuclei of atoms their great stability. The range of the strong force extends only as far as the nucleus (10^{-15}m).

Electromagnetic force – This is responsible for the forces between charged particles such as protons and electrons, so can be attractive or repulsive. It's also responsible for electromagnetic emission and absorption, and it controls atomic structure, which allows atoms to bind together to make molecules.

TRANSFERS

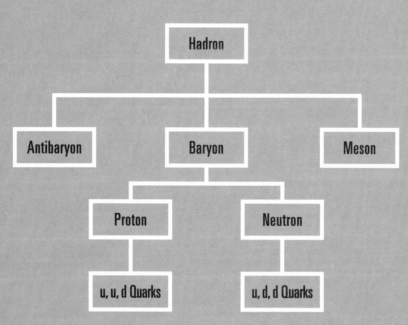

THE HADRON FAMILY
A partial family tree of the Hadron family showing where protons and neutrons, which make up the nucleus of an atom, fit in.

The weak force – Acting at a shorter range than even the strong force, the weak force interaction becomes particularly important during nuclear reactions, where it can switch the flavour of a quark, thereby changing a nuclear particle from one type to another. One example of this is beta decay, in which a neutrally charged neutron is converted into a positively charged proton, with a high-energy (and negatively charged) electron then emitted. Another example is a star's nuclear fusion reactions, which see the production of deuterium and then helium from hydrogen.

Gravitational force – Here's the force that we're all acutely aware of in our daily lives. It's the cause of stuff falling to the ground and of the Moon orbiting the Earth. Gravity is only a noticeable force at large masses – at the atomic level it's negligible due to its strength being 10^{40} times weaker than the strong force. The range of gravitational force, however, is infinite.

THE MISSING PIECE OF THE PUZZLE

As promised earlier, we're now ready to tackle what gauge bosons are. In the Standard Model, these are force carrier particles. They're responsible for three of the four fundamental forces (the Standard Model's main failing is that it can't actually explain gravity). Gluons carry the strong force; photons carry the electromagnetic force; and W and Z bosons lie behind the weak force.

The Higgs boson, the final piece in the Standard Model jigsaw, was proposed the same year as the quark model, in 1964. It, along with its connected Higgs field, has played a key role in the development of Standard Model by explaining how each of the elementary particles come to have mass and why photons and gluons don't. For a long time, it was the only particle in the model yet to be observed. However, on 4 July 2012, two separate teams of researchers at the Large Hadron Collider (LHC) at CERN (European Council for Nuclear Research) in Switzerland presented experimental results showing that, after some rather committed searching, they had almost certainly found evidence of a Higgs boson-like particle. More data and further analysis followed, and CERN finally referred to the particle as 'the Higgs boson' in July 2017.

A Key to Unlocking the World

THE MODERN PERIODIC TABLE OF ELEMENTS

This is the version of the periodic table that features most commonly in books and on classroom walls. It may look boring, but it's an incredibly powerful tool that encodes a lot of information to the knowing student. The table is ordered by atomic number, which refers to the number of protons in an element's nucleus. For example, all carbon atoms have six protons in their nucleus. The columns in the periodic table are called groups. The rows are called periods. So, potassium is in group 1, period 4. Elements in a group have similar chemical properties – for example, lithium and sodium both react with water to produce a metal hydroxide (an alkali) and hydrogen.

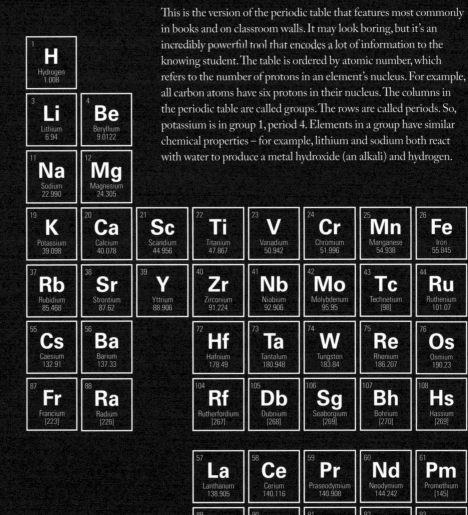

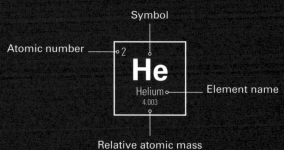

Atomic number — 2

Symbol

He

Helium — Element name

4.003

Relative atomic mass

						2 **He** Helium 4.003
	5 **B** Boron 10.81	6 **C** Carbon 12.011	7 **N** Nitrogen 14.007	8 **O** Oxygen 15.999	9 **F** Fluorine 18.998	10 **Ne** Neon 20.180
	13 **Al** Aluminium 26.982	14 **Si** Silicon 28.085	15 **P** Phosphorus 30.974	16 **S** Sulphur 32.06	17 **Cl** Chlorine 35.45	18 **Ar** Argon 39.948

27 **Co** Cobalt 58.933	28 **Ni** Nickel 58.693	29 **Cu** Copper 63.546	30 **Zn** Zinc 65.38	31 **Ga** Gallium 69.723	32 **Ge** Germanium 72.630	33 **As** Arsenic 74.922	34 **Se** Selenium 78.971	35 **Br** Bromine 79.904	36 **Kr** Krypton 83.798
45 **Rh** Rhodium 102.906	46 **Pd** Palladium 106.42	47 **Ag** Silver 107.868	48 **Cd** Cadmium 112.414	49 **In** Indium 114.818	50 **Sn** Tin 118.710	51 **Sb** Antimony 121.760	52 **Te** Tellurium 127.60	53 **I** Iodine 126.904	54 **Xe** Xenon 131.293
77 **Ir** Iridium 192.217	78 **Pt** Platinum 195.084	79 **Au** Gold 196.967	80 **Hg** Mercury 200.592	81 **Tl** Thallium 204.38	82 **Pb** Lead 207.2	83 **Bi** Bismuth 208.980	84 **Po** Polonium [209]	85 **At** Astatine [210]	86 **Rn** Radon [222]
109 **Mt** Meitnerium [278]	110 **Ds** Darmstadtium [281]	111 **Rg** Roentgenium [280]	112 **Cn** Copernicium [285]	113 **Nh** Nihonium [286]	114 **Fl** Flerovium [289]	115 **Mc** Moscovium [289]	116 **Lv** Livermorium [293]	117 **Ts** Tennessine [294]	118 **Og** Oganesson [294]

62 **Sm** Samarium 150.36	63 **Eu** Europium 151.964	64 **Gd** Gadolinium 157.25	65 **Tb** Terbium 158.925	66 **Dy** Dysprosium 162.500	67 **Ho** Holmium 164.930	68 **Er** Erbium 167.259	69 **Tm** Thulium 168.934	70 **Yb** Ytterbium 173.045	71 **Lu** Lutetium 174.967
94 **Pu** Plutonium [244]	95 **Am** Americium [243]	96 **Cm** Curium [247]	97 **Bk** Berkelium [247]	98 **Cf** Californium [251]	99 **Es** Einsteinium [252]	100 **Fm** Fermium [257]	101 **Md** Mendelevium [258]	102 **No** Nobelium [259]	103 **Lr** Lawrencium [262]

The Periodic Snail

ORDERING THE ELEMENTS

Everything on Earth is made from one or more of the 118 chemical elements shown in the periodic table (see previous). The version of the table that we use today was first proposed in 1869 by the Russian chemist Dimitri Mendeleev, who wanted to illustrate recurring trends in the properties of the different elements. He put them into a systematic order, and this is what makes the periodic table such a powerful tool. Knowing the order allows us to predict how elements will behave, and possibly combine, to make new substances.

In Mendeleev's time, only about sixty elements were known to science. The Russian left spaces in his table predicting new elements, and as these and others have been discovered, the layout of the table has been refined and extended. Of the known elements, ninety-four occur naturally on Earth. The rest are human-made.

As with great works of literature, the periodic table can be appreciated on many levels, from when it's simply a useful system of categorisation during our school years to its subtlety and nuance as our scientific knowledge grows. So, it might not come as too much of a surprise to learn that ever since it was created by Mendeleev, the periodic table has been presented in literally hundreds of forms, not just the one on the previous page. A particular theme within many of these alternative tables has been the use of a spiral. Examples include artist Edgar Longman's mural from the 1951 Festival of Britain Exhibition of Science and plant scientist Philip Stewart's *Chemical Galaxy II*, which is presented with an interstellar look and can be seen at www.chemicalgalaxy.co.uk. One of the best is German-American chemist Theodor Benfey's 'periodic snail', which he developed 'to emphasise the complex yet beautiful periodicity in the properties of the chemical elements'.

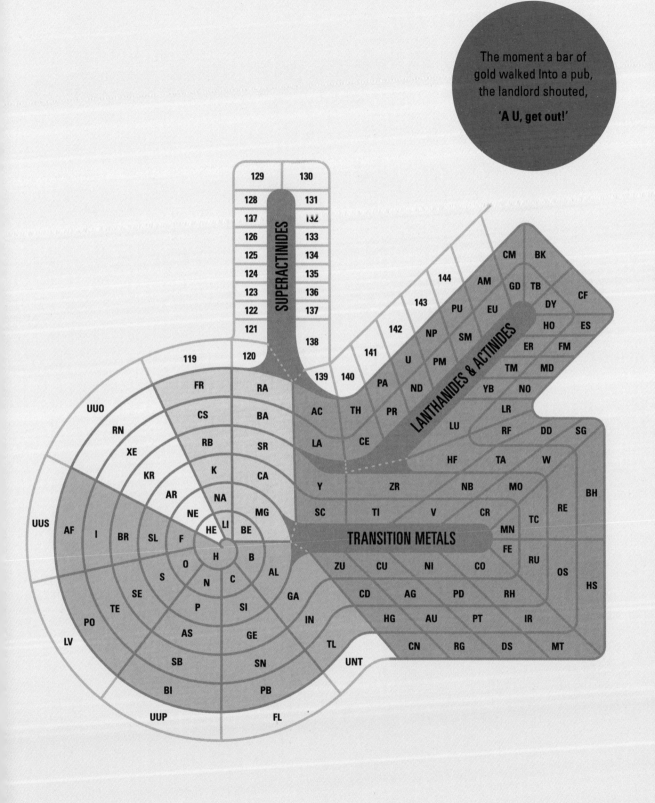

The moment a bar of gold walked Into a pub, the landlord shouted,

'A U, get out!'

A Singing Science Sensation

THE LARGE HADRON COLLIDER RAP

In the build-up to the official switch-on of the LHC at CERN, some scientists working there took it upon themselves to write, record and perform a rap to show the world what their research was all about. Within a week of it being posted on YouTube, more than two million people had viewed it – and at the time of writing the figure stands at more than eight million.

With lyrics by science writer Kate McAlpine and music by Will Barras, a linguistics PhD, and featuring a variety of anonymous dancing scientists in numerous locations at CERN, the song is a brilliant piece of science communication (search for 'LHC rap' on YouTube to watch the video). With special thanks to Kate for allowing me to reproduce her lyrics:

Artist impression of particle traces from a LHC collision.

Twenty-seven kilometres of tunnel under ground
Designed with mind to send protons around
A circle that crosses through Switzerland and France
Sixty nations contribute to scientific advance
Two beams of protons swing round, through the ring
*   they ride*
'Til in the hearts of the detectors, they're made to collide
And all that energy packed in such a tiny bit of room
Becomes mass, particles created from the vacuum
And then …

Chorus:

LHCb sees where the antimatter's gone
ALICE looks at collisions of lead ions
CMS and ATLAS are two of a kind
They're looking for whatever new particles
*   they can find.*
The LHC accelerates the protons and the lead
And the things that it discovers will rock you
*   in the head.*

We see asteroids and planets, stars galore
We know a black hole resides at each galaxy's core
But even all that matter cannot explain
What holds all these stars together –
*   something else remains*
This dark matter interacts only through gravity
And how do you catch a particle there's no
*   way to see*
Take it back to the conservation of energy
And the particles appear, clear as can be

You see particles flying, in jets they spray
But you notice there ain't nothin', goin' the other way
You say, 'My law has just been violated – it don't
*   make sense!*
There's gotta be another particle to make
*   this balance.'*

And it might be dark matter, and for first
Time we catch a glimpse of what must fill most
    of the known 'verse.
Because …

[Chorus]

Antimatter is sort of like matter's evil twin
Because except for charge and handedness of spin
They're the same for a particle and its anti-self
But you can't store an antiparticle on any shelf
Cuz when it meets its normal twin, they
    both annihilate
Matter turns to energy and then it dissipates

When matter is created from energy
Which is exactly what they'll do in the LHC
You get matter and antimatter in equal parts
And they try to take that back to when the
    universe starts
The Big Bang – back when the matter all exploded
But the amount of antimatter was somehow eroded
Because when we look around we see that
    matter abounds
But antimatter's nowhere to be found.
That's why …

[Chorus]

The Higgs boson – that's the one that everybody
    talks about
And it's the one sure thing that this machine will
    sort out
If the Higgs exists, they ought to see it right away
And if it doesn't, then the scientists will finally say
'There is no Higgs! We need new physics to account
    for why
Things have mass. Something in our Standard
    Model went awry.'

But the Higgs – I still haven't said just what it does
They suppose that particles have mass because
There is this Higgs field that extends through
    all space
And some particles slow down while other
    particles race
Straight through like the photon – it has no mass
But something heavy like the top quark, it's draggin'
    its a**
And the Higgs is a boson that carries a force
And makes particles take orders from the field that
    is its source.
They'll detect it …

[Chorus]

Now some of you may think that gravity is strong
Cuz when you fall off your bicycle it don't take long
Until you hit the earth, and you say, 'Dang, that hurt!'
But if you think that force is powerful, you're wrong.
You see, gravity – it's weaker than Weak
And the reason why is something many scientists seek
They think about dimensions – we just live in three
But maybe there are some others that are too small to see
It's into these dimensions that gravity extends
Which makes it seem weaker, here on our end.
And these dimensions are 'rolled up' – curled so tight
That they don't affect you in your day-to-day life
But if you were as tiny as a graviton
You could enter these dimensions and go wandering on
And they'd find you …

[Chorus]

Light-years Away

WHY IS THE SKY DARK AT NIGHT?

Does this sound like a strange question? Well, it shouldn't. Named after the German astronomer Heinrich Wilhelm Olbers, who raised it in 1823 (even though German astronomer Johannes Kepler was the first to do so, in 1610), Olbers's paradox asks why the night sky isn't uniformly bright, given, as the American writer Edgar Allen Poe wonderfully put it in his prose poem *Eureka*, published in 1848:

Were the succession of stars endless, then the background of the sky would present us a uniform luminosity, like that displayed by the Galaxy — since there could be absolutely no point, in all that background, at which would not exist a star.

In fact, the suggested explanation that Poe went on to give for this was very close to the one given more than fifty years later by Scottish-Irish engineer and physicist Lord Kelvin, which is still generally accepted. As Poe says:

The only mode, therefore, in which we could comprehend the voids which our telescopes find in innumerable directions, would be by supposing the distance of the invisible background so immense that no ray from it has yet been able to reach us at all.

In other words, light from a great majority of stars in the universe hasn't yet reached us.

Unweaving the Rainbow

In science, one tries to tell people, in such a way as to be understood by everyone, something that no one ever knew before. But in the case of poetry, it's the exact opposite!

PAUL DIRAC (1902–1984), ENGLISH PHYSICIST

When I heard the learn'd astronomer,
When the proofs, the figures, were ranged in
*   columns before me,*
When I was shown the charts and diagrams,
*   to add, divide, and measure them,*
When I sitting heard the astronomer where he
*   lectured with much applause in the lecture-room,*
How soon unaccountable I became tired and sick,
Till rising and gliding out I wander'd off by myself,
In the mystical moist night-air, and from
*   time to time,*
Look'd up in perfect silence at the stars.

THE BEAUTY IN SCIENCE

Many poets have railed against what they perceive to be the reductive nature of science. John Keats certainly didn't pull any punches when, in his 1820 poem 'Lamia', he referred to science as a 'cold philosophy' that would 'unweave a rainbow', thereby stripping the world of its beauty. But perhaps the most eloquent expression of this ungenerous view of science can be found in American poet Walt Whitman's deceptively simple 'When I Heard the Learn'd Astronomer', published in 1865:

The American physicist and bongo player Richard Feynman produced the perfect riposte to this view when he said in his 1960s' lecture 'The Relation of Physics to other Sciences':

Poets say science takes away from the beauty of the stars – mere globs of gas atoms. Nothing is 'mere'. I too can see the stars on a desert night, and feel them. But do I see less or more? The vastness of the heavens stretches my imagination – stuck on this carousel my little eye can catch one-million-year-old light … It does not do harm to the mystery to know a little about it. For far more marvellous is the truth than any artists of the past imagined!

The New Atlantis

AN ADMINISTRATION OF LEARNING

It might not come as a surprise to hear the roots of the Royal Society in London may lie in the work of English philosopher Francis Bacon (1561–1626). He is, after all, considered by many to be 'the father of experimental science'. But this belies the remarkable nature of this story.

According to the 17th-century English writer John Aubrey, William Harvey – physician to kings James I and Charles I – once derisively said of his patient Francis Bacon that he wrote philosophy 'like a Lord Chancellor'. That Bacon was, in fact, Lord Chancellor was part of the point. But it was also a criticism of a bureaucratic style of

Francis Bacon

writing that Harvey felt Bacon had. While Harvey was clearly of the view that Bacon should have just concentrated on the day job, to Bacon his career was part of what provided him with the mental tools for the task he had in mind – to create an 'administration of learning'.

In *New Atlantis*, a fictional utopian work published after his death, Bacon describes an institution called Salomon's House 'dedicated to the study of the works and creatures of God', the end of which is 'the knowledge of causes, and secret motions of things; and the enlarging of the bounds of human empire, to the effecting of all things possible'. The brethren of the house had access to a great deal of natural and artificial resources, enabling pretty much every area imaginable to be studied; there was a huge number of staff all contained within a highly structured hierarchy, from 'servants and attendants' to 'novices and apprentices', to the particular specialised roles of the brethren themselves, including 'Merchants of Light', who travelled to other nations to retrieve their knowledge, and 'Pioneers', whose role was to devise and 'try new experiments'.

Bacon was essentially writing a manifesto for what he saw as the perfect scientific research establishment, the apotheosis of the systemisation of knowledge and learning. This was not a work for the general reader. But it didn't need to be. The Prussian writer

EXALTED COMPANY

The frontispiece of Thomas Sprat's *The History of the Royal Society of London* (1667) showing, from left to right, the Society's first president, William Brouncker, King Charles II and Francis Bacon.

Benevolent maunderers
stand up and say
That black and white are but
extremes of grey;
Stir up the black creed with the white,
The grey they make will be just right.

**WRITTEN BY T. H. HUXLEY, APPARENTLY IN
RESPONSE TO A CHURCH CONGRESS**

A CENTRE OF LEARNING
Burlington House, home to the Society
between 1873 and 1967.

Samuel Hartlib, who spent most of his intellectual life in England, proved an influential advocate of Bacon's ideas in the mid-17th century. Hartlib believed passionately in educational reform and felt this could be best achieved via the real-life establishment of Salomon's House. He even went as far as identifying the Chelsea College, a divinity college, as its possible location and lobbied Parliament to raise money for the project – but to no avail. However, among Hartlib's close circle of intellectual friends were the English political economist William Petty and the Anglo-Irish scientist Robert Boyle, who championed the idea of a 'philosophical college' that took 'the whole body of mankind for their care'. Both Petty and Boyle would be among the twelve people present at the inaugural meeting of the Royal Society in 1660, less than forty years after Bacon's death.

RAISING THE PUBLIC PROFILE

The first few years of the Society weren't easy – not everyone was convinced of its value and principles. Rather bizarrely for an organisation barely three years old, an official history by one of its fellows was commissioned to help present its credentials. Published in 1667, Thomas Sprat's *The History of the Royal Society of London* featured Francis Bacon in its frontispiece along with King Charles II and William Brouncker – a mathematician and the Society's first president. Through this signalling, and the language used inside the book, it was clear how Baconian the Royal Society wanted to appear. The implication was clear – the utopian Salomon's House was now a reality.

Approval wasn't unanimous, and the Society still had to suffer its detractors. The Anglo-Irish writer Jonathan Swift lampooned it superbly in his description of the grand academy of Lagado in his book *Gulliver's Travels*, published in 1726:

> *This academy is not an entire single building, but a continuation of several houses on both sides of a street.... Every room has in it one or more projectors; and I believe I could not be in fewer than five hundred rooms.*
>
> *The first man I saw was of a meagre aspect, with sooty hands and face, his hair and beard long, ragged, and singed in several places. His clothes, shirt, and skin were all of the same colour. He has been eight years upon a project for extracting sunbeams out of cucumbers, which were to be put in phials hermetically sealed, and let out to warm the air in raw inclement summers. He told me, he did not doubt, that, in eight years more, he should be able to supply the governor's gardens with sunshine, at a reasonable rate....*

During its 360-year history, the Royal Society has had some truly remarkable people as its presidents, many of whom have embodied the society's motto '*Nullius in verba*' (roughly translated as 'take nobody's word for it'). These have included Christopher Wren, who was one of the twelve founding members, Samuel Pepys, Isaac Newton, Hans Sloane, Joseph Banks, Humphry Davy, T. H. Huxley, Joseph Lister, Ernest Rutherford and Howard Florey. In 1756, Benjamin Franklin was the first American elected as a fellow of the society. In many respects, however, it was still a mirror of society at large. It seems incredible today to learn that it wasn't until 1945 that the first two female fellows of the Royal Society were elected – the Irish crystallographer Kathleen Lonsdale and the English biochemist Marjory Stephenson.

A World of Worlds in Every Earring

A WISE AND LEARNED LADY

Margaret Cavendish (c. 1623–1673), Duchess of Newcastle upon Tyne, was formerly maid of honour to the wife of King Charles I, Queen Henrietta Maria, whom she accompanied into exile in Paris in 1644. There she met, among others, the French philosopher René Descartes – and her future husband, William Cavendish.

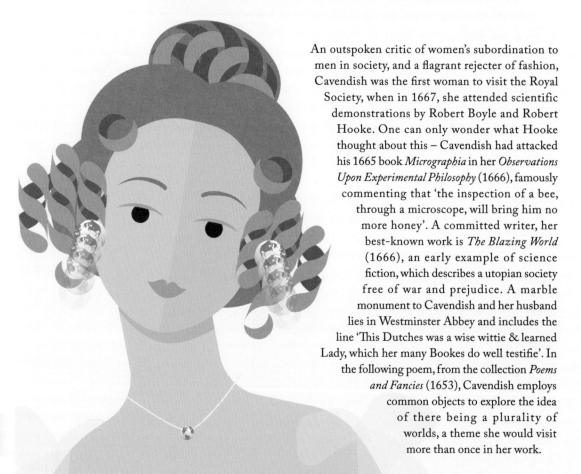

An outspoken critic of women's subordination to men in society, and a flagrant rejecter of fashion, Cavendish was the first woman to visit the Royal Society, when in 1667, she attended scientific demonstrations by Robert Boyle and Robert Hooke. One can only wonder what Hooke thought about this – Cavendish had attacked his 1665 book *Micrographia* in her *Observations Upon Experimental Philosophy* (1666), famously commenting that 'the inspection of a bee, through a microscope, will bring him no more honey'. A committed writer, her best-known work is *The Blazing World* (1666), an early example of science fiction, which describes a utopian society free of war and prejudice. A marble monument to Cavendish and her husband lies in Westminster Abbey and includes the line 'This Dutches was a wise wittie & learned Lady, which her many Bookes do well testifie'. In the following poem, from the collection *Poems and Fancies* (1653), Cavendish employs common objects to explore the idea of there being a plurality of worlds, a theme she would visit more than once in her work.

'OF MANY WORLDS IN THIS WORLD'

Just like as in a Nest of Boxes round,
Degrees of Sizes in each Box are found:
So, in this World, may many others be
Thinner and less, and less still by degree:
Although they are not subject to our sense,
A World may be no bigger than Two-pence.
NATURE is curious, and such Works may shape,
Which our dull senses easily escape:
For Creatures, small as Atoms, may be there,
If every one a Creature's Figure bear.
If Atoms Four, a World can make, then see
What several Worlds might in an Ear-ring be:
For, Millions of those Atoms may be in
The Head of one small, little, single Pin.
And if thus small, then Ladies may well wear
A World of Worlds, as Pendants in each Ear.

Twenty years after the poem was published, the Dutch scientist Antonie van Leeuwenhoek sent letters to the Royal Society describing his experiments with the vastly improved microscopes he had built. Over the following years, he recorded many original observations, including seeing what he termed 'animalcules' (bacteria) in plaque scraped from his teeth. An irony perhaps was that the inspiration for Leeuwenhoek's work had come from Hooke's *Micrographia*, which Cavendish had disparaged. Well, nobody's perfect.

Science was for Marx a historically dynamic, revolutionary force. However great the joy with which he welcomed a new discovery in some theoretical science whose practical application perhaps it was as yet quite impossible to envisage, he experienced quite another kind of joy when the discovery involved immediate revolutionary changes in industry and in historical development in general.

FRIEDRICH ENGELS (GERMAN PHILOSOPHER, 1820–1895), FROM HIS EULOGY AT THE FUNERAL OF KARL MARX, 1883

Scratching at the Surface

Fluorite

Topaz

Quartz

THE RELATIVE HARDNESS OF BEING

Devised by the German geologist
Friedrich Mohs (1773–1839), the Mohs scale
of mineral hardness arranges ten minerals
in an order of increasing hardness. The key
to the scale is that each mineral can be scratched
by any below it but none above it.

The scale does not simply apply to these ten minerals – it can
be used to assess the hardness of any solid. So, if a solid can be
scratched by apatite but not fluorite, then its Mohs hardness
is between 4 and 5. Graphite (found in pencil lead) has a
Mohs hardness of between 1 and 2. Anything up to 2.5 can
typically be scratched by a fingernail, up to 4 by a coin and up
to 6 by a knife. The scale is a relative one – the difference in
absolute hardness between each mineral isn't consistent.

Mohs hardness	Mineral
1 (softest)	Talc
2	Gypsum
3	Calcite
4	Fluorite
5	Apatite
6	Feldspar
7	Quartz
8	Topaz
9	Corundum
10 (hardest)	Diamond

A Law by Any Other Name ...

CONTESTED CLAIMS

If I had to give three names synonymous with scientific laws, I would likely respond Boyle, Hooke and Newton. The first of these, Anglo-Irish scientist Robert Boyle (1627–1691), a founding member of the Royal Society, gave his name to Boyle's law, which concerns the relationship between the pressure and volume of a gas, as given by the equation:

$$pV = \text{constant}$$

On the Continent, however, the law is known by a different name, Mariotte's law, after the French priest and scientist Edmé Mariotte, who proposed his own version of the law. He was responsible for a number of other discoveries, including the fact that the place where the optic nerve joins the eye's retina is a blind spot – something you rarely realise because your brain fills in the missing area, backed up by information from your other eye. It's not hard to find areas of disagreement between the French and the English, but in this instance, the English-speaking world is right – priority lies with Boyle, who published his discovery more than ten years before Mariotte.

The priest also lost out somewhere else. In 1967, English actor Simon Prebble convinced the London store Harrods to stock a device that he called 'Newton's cradle', which became a popular 'executive toy'. In fact, it was Mariotte who had first conducted experiments exploring the phenomena demonstrated by the collision of pendulum balls. In 1671, Mariotte presented his findings to the French Academy of Sciences and two years later published them. Newton acknowledged Mariotte's work fourteen years later in his *Principia* when concluding that momentum, which Newton called 'quantity of motion', is always conserved. Whether an executive toy named 'Mariotte's cradle' would have sold quite so well is open to conjecture.

Father of Chemistry and Uncle of the Earl of Cork

REPORTED EPITAPH OF ROBERT BOYLE

Just Not My Cup of Tea

BOILING POINTS

In his book *Inflight Science*, popular science writer Brian Clegg discusses whether it's possible to get a decent cup of tea while on an aeroplane. His answer is an emphatic 'no', for the simple reason that the cabin crew cannot boil water at a tea connoisseur's ideal of 100°C (212°F), because the air pressure in the cabin is lower than we generally experience in everyday life. Since the boiling point of water (or any substance) drops in accordance with pressure, water heated in an aircraft cabin will reach boiling point before it gets to 100°C (it will boil at 93°C, or 199°F).

Tea isn't the only comestible that can have its quality affected by the differences in pressure caused by altitude, as Charles Darwin discovered in 1835 during his voyage on HMS *Beagle*:

Having crossed the Peuquenes [a Chilean mountain ridge], we descended into a mountainous country … The elevation was probably not under 11,000 feet [3,350m] … At the place where we slept water necessarily boiled, from the diminished pressure of the atmosphere, at a lower temperature than it does in a less lofty country … .
Hence the potatoes, after remaining for some hours in the boiling water, were nearly as hard as ever. The pot was left on the fire all night, and next morning it was boiled again, but yet the potatoes were not cooked.

Darwin's party would have found their water boiling at a temperature of approximately 88.7°C. As a nation of tea drinkers, the British are lucky that their highest point of altitude is two-fifths that experienced by Darwin on the expedition. But, so they're forewarned, the following table shows how the boiling point of water varies between various locations around the world.

	Altitude	Boiling point of water
Dead Sea	−423m	101.4°C
Sea-level	0m	100°C
Ben Nevis	1,344m	95.6°C
Cabin pressure of Boeing 767 at cruising altitude	2,100m	93°C
Mexico City	2,240m	92.6°C
Mont Blanc	4,810m	83.5°C
Kilimanjaro	5,895m	79.5°C
Everest	8,848m	68.0°C

Number Sifting Made Simple

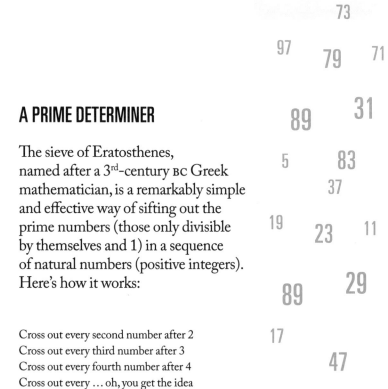

A PRIME DETERMINER

The sieve of Eratosthenes, named after a 3rd-century BC Greek mathematician, is a remarkably simple and effective way of sifting out the prime numbers (those only divisible by themselves and 1) in a sequence of natural numbers (positive integers). Here's how it works:

Cross out every second number after 2
Cross out every third number after 3
Cross out every fourth number after 4
Cross out every … oh, you get the idea

This is what the 1 to 100 look like after doing this.

You may notice something rather interesting about the numbers remaining – they often come in pairs, differing only by 2, for example 3/5, 11/13, 17/19, 29/31 and so on. Calculations suggest that there are always such pairs, known as twin primes, although so far no one has provided a proof of this.

The Prussian mathematician Christian Goldbach wrote a letter to his Swiss counterpart Leonhard Euler in 1742 conjecturing that every even integer greater than 2 can be stated as the sum of two primes. For example, $8 = 3 + 5$, $18 = 11 + 7$ and so on. No exception to the conjecture has so far been found, but, as with twin primes, no proof has been given either. Bloomsbury and Faber, the American and British publishers respectively of the novel *Uncle Petros and Goldbach's Conjecture*, by the Greek writer Apostolos Doxiadis, announced in 2000 that they would pay a prize of $1 million to anyone who proved the conjecture within two years of the book's publication. No one did and the publishers' bank managers were delighted.

The Body Electric

FRANKENSTEIN AND THE REANIMATING POWER OF ELECTRICITY

It was on a dreary night of November that I beheld the accomplishment of my toils. With an anxiety that almost amounted to agony, I collected the instruments of life around me, that I might infuse a spark of being into the lifeless thing that lay at my feet. It was already one in the morning; the rain pattered dismally against the panes, and my candle was nearly burnt out, when, by the glimmer of the half-extinguished light, I saw the dull yellow eye of the creature open; it breathed hard, and a convulsive motion agitated its limbs.

Opening of Chapter V of *Frankenstein* by Mary Shelley, 1818

Mary Wollstonecraft Godwin – the future Mary Shelley – was born on 30 August 1797 in Somers Town, between Camden and St Pancras in London. Both her parents were radical writers. Her mother, Mary Wollstonecraft, was the author of *A Vindication of the Rights of Woman* (1792), which argued for the equal education of both boys and girls, among other things. Her father was William Godwin, an atheist and an early proponent of anarchist principles, who wrote numerous books, including *Enquiry Concerning Political Science* (1793) and the novel *Caleb Williams* (1794).

Mary's mother died of puerperal fever, a disease almost certainly accidentally introduced by her doctor during childbirth, when Mary was just eleven days old. Her father remarried, and Mary was brought up in a house full of books and frequented by many great thinkers, including the poet and polymath Samuel Taylor Coleridge. Godwin described the fifteen-year-old Mary as 'singularly bold, somewhat imperious, and active of mind. Her desire for knowledge is great, and her perseverance in everything she undertakes almost invincible'.

In 1814, a few months before her seventeenth birthday, Mary and the twenty-one-year-old poet Percy Bysshe Shelley fell in love. Her father disapproved, and they eloped to France, eventually marrying at the end of 1816 in London while *Frankenstein*, widely regarded as the 'first true work of science fiction', was being written.

Death already haunted nineteen-year-old Mary Shelley's life when she began to write *Frankenstein* in the summer of 1816. Her birth had been closely followed by the loss of her mother, and in 1815 Mary's first child died after only a few weeks. She later wrote about a recurring dream in which 'my little baby came to life again; that it had only been cold, and that we rubbed it before the fire, and it lived'.

In June 1816, Mary, Percy Shelley and Mary's stepsister, Claire Clairmont, lived close to the poet Lord Byron and his physician Dr John William Polidori near Geneva, Switzerland. One evening that month the group launched a ghost story contest. Recalling events fifteen years later, Mary wrote:

> *Lord Byron and Shelley … discussed … the nature of the principle of life, and whether there was any probability of its ever being discovered and communicated … Perhaps a corpse would be reanimated; galvanism had given token of such things; perhaps the component parts of a creature might be manufactured, brought together, and endued with vital warmth.*
>
> *Night waned upon this talk, and even the witching hour had gone by before we retired to rest. When I placed my head on my pillow, I did not sleep, nor could I be said to think. My imagination, unbidden, possessed and guided me, gifting the successive images that arose in my mind with a vividness far beyond the usual bound of reverie. I saw – with shut eyes, but acute mental vision – I saw the pale student of unhallowed arts kneeling beside the thing he had put together; I saw the hideous phantasm of a man stretched out; and then, on the working of some powerful engine, show signs of life, and stir with an uneasy, half-vital motion.*

Nearly forty years before Mary Shelley wrote *Frankenstein*, the Italian anatomist Luigi Galvani had spent several years dissecting frogs when, in 1780, his research interests took a dramatic turn. As his metal scalpel touched the inner nerve of a frog's leg, violent contractions were suddenly observed. Similar results were seen when the steel scalpel touched a brass hook used to hold a leg in place. Galvani went on to perform numerous experiments, including one in which stormy weather conditions produced similar muscle contractions. These results, along with a series of follow-up investigations, led Galvani to confirm and extend the contemporary theory of animal electricity, and 1791 saw the publication of his *Commentary on the Effects of Electricity on Muscular Motion*.

Twelve years later, in 1803, Galvani's nephew Giovanni Aldini performed several experiments on the body of George Foster, a very recently executed London murderer. As reported at the time, 'On the first application of the arcs, the jaw began to quiver, the adjoining muscles were horribly contorted, and the left eye actually opened'. Investigations continued. 'The conductors being applied to the ear, and to the rectum, excited in the muscles contractions much stronger than in the preceding experiments … to [almost] give an appearance of reanimation'.

Of the only twelve copies of the first edition of Galvani's *Commentary*, one was sent to Italian physicist Alessandro Volta. At first, Volta was an enthusiastic supporter of Galvani's conclusions. However, a change came when he started to look carefully at his compatriot's experiments, repeating many of them himself. Where Galvani saw the generation of the electric current as occurring in the frog's muscles, Volta's brilliance was to realize that the presence of two different metals was the cause of the current. Volta believed that the frog essentially acted as an extremely sensitive electrometer (an instrument for measuring electric current). So, he began to investigate how he could increase the current produced by two different metals.

Volta's experiments led to his invention of the electric pile, which we would call a battery (a battery is really a series of such piles or cells). After experimenting with various pairs of metals, Volta settled on zinc and silver as providing the greatest effect. However, he also recognised that Galvani's frogs had contributed something essential to the process – a conductive liquid. So, Volta's first electric pile consisted of alternating zinc and silver discs, with salt-water-soaked cardboard between each disc and wires connected at both ends. The size of the current depended on the metals used and could be increased by adding more discs. This was the first time a steady current could be produced, and it revolutionised science.

The Royal Society in London announced Volta's discovery of his 'electric pile' in 1800 – and almost immediately, the English chemist Humphry Davy began investigating its potential. In the space of two years (1807–1808), and using an electric battery, he discovered and named the elements potassium, sodium, calcium and barium.

While writing Frankenstein during the autumn of 1816, Mary Shelley read Davy's *Elements of Chemical Philosophy* (1812), featuring his work with chemicals and electricity. Some academics view the work of Davy as providing Frankenstein's model of the scientist as a heroic Romantic adventurer. Science fiction indeed.

The American academic Charles E. Robinson has stated that as many as eleven texts of *Frankenstein* have existed. The earliest that we have access to is the first complete draft of the novel, of which about 87 per cent survives. As Mary wrote the manuscript, Percy Shelley made various editorial suggestions and changes. This included deleting 'And' and 'But' from the start of some of Mary's sentences and paragraphs and changing various instances of 'that' to 'which' to introduce relative clauses. In total, it's estimated that Percy contributed 4,000 to 5,000 words of the final 72,000-word novel. It remains very much Mary's masterpiece, however.

> *To examine the causes of life, we must first have recourse to death.*
>
> **FROM CHAPTER III OF**
> ***FRANKENSTEIN***
> **BY MARY SHELLEY, 1818**

For Goodness' Sake, Show Your Working

THE LITTLE AND LAST THEOREMS

Pierre de Fermat (1601–1665) was a lawyer and amateur mathematician – amateur because he wasn't always interested in providing proofs for his theorems, as both of the examples of his work we are about to look at show. He spent most of his adult life in the French city of Toulouse and is perhaps most famous for his little and last theorems.

Fermat's little theorem, which can be expressed in various ways, is a method for finding out whether a number, p, is prime or not. If it is, and a is any positive integer, then

$$a^{p-1} - 1$$

will be divisible by p.
For example, if we let p = 13 and a = 2, then

$$2^{12} - 1 = 4{,}095$$

which is divisible by 13 (when 4,095 is divided by 13, the answer we get is 315), so 13 is a prime number.

As was typical of Fermat, when he sent his theorem to a friend, he provided no proof because he feared it was too long. It would be another 100 years before mathematician Leonhard Euler published a proof.

And it took more than 350 years for a proof to be found to an assertion Fermat wrote in the margins of his copy of *Arithmetica*, by the 3rd-century AD Greek mathematician Diophantus, and which became known as Fermat's last theorem ('To divide a cube into two other cubes, a fourth power or in general any power whatever into two powers of the same denomination above the second is impossible, and I have assuredly found an admirable proof of this, but the margin is too narrow to contain it'). In essence, it stated that no positive integers could satisfy the following equations:

$$x^n + y^n = z^n, \text{ where } n \geq 3$$

The special case of Fermat's last theorem is Pythagoras's theorem, when n = 2. Here, 3, 4 and 5 are possible solutions.

The problem perplexed mathematicians for centuries. Then, in June 1993, the English mathematician Andrew Wiles announced he had found a proof (in a document running to 200 pages). But within a few months an error was spotted. Wiles tried to fix it with the help of another English mathematician, Richard Taylor. A year later, two papers were produced – a long one by Wiles and a short one by Wiles and Taylor. This time nobody disputed the proof, and the story was finally brought to a close. The prestigious journal *Annals of Mathematics* gave over its entire May 1995 issue to the proof.

I'm very well acquainted, too, with matters mathematical, I understand equations, both the simple and quadratical, About binomial theorem I'm teeming with a lot o' news, With many cheerful facts about the square of the hypotenuse. I'm very good at integral and differential calculus; I know the scientific names of beings animalculous: In short, in matters vegetable, animal, and mineral, I am the very model of a modern Major-General.

**W. S. GILBERT (1836–1911),
ENGLISH DRAMATIST AND LIBRETTIST
'I AM THE VERY MODEL OF A MODERN
MAJOR-GENERAL', FROM *THE PIRATES
OF PENZANCE* (1879)**

Kicking Up a Stink

FARADAY'S SMELLY THAMES TRAVERSAL

The River Thames played a prominent role in the opening
ceremony of the London 2012 Olympics – who can forget
the iconic image of footballer David Beckham, winner of 115
England caps, zooming up the river in a speedboat carrying the
Olympic torch? Leaving aside anachronism, this would have
been virtually impossible to do 160 years ago because of the
overwhelming smell that Beckham would have experienced due
to human contamination (I'm being polite here) of the river –
it's hard to look dashing when you're suppressing the desire
to retch – I know, I've tried.

The summer of 1858 was particularly hot and the resulting smell so bad that
it was labelled The Great Stink. No class of person was immune to its charms,
and it eventually led to a new sewage system, devised by the civil engineer Joseph
Bazalgette. However, that was still three years away when, in 1855, the eminent
scientist Michael Faraday felt compelled to write to the editor of *The Times* regarding
the state of the Thames (see opposite). His letter paints an unpleasantly vivid picture
of the polluted state of the river at that time.

Two weeks after Faraday's letter was printed, *Punch* magazine published a witty
caricature of 'Faraday giving his card to Father Thames', adding, 'And we hope the
Dirty Fellow will consult the learned Professor'.

Sir,

I traversed this day by steamboat to the space between London and Hungerford bridges between half past one and two o'clock; it was low water, and I think the tide must have been near the turn. The appearance and the smell of the water forced themselves at once to my attention. The whole of the river was an opaque pale brown fluid. In order to test the degree of opacity, ` I tore up some white cards into pieces, moistened them so as to make them sink easily below the surface, and then dropped some of these pieces into the water at every pier the boat came to; before they had sunk an inch below the surface they were indistinguishable, though the sun shone brightly at the time; and when the pieces fell edgeways the lower part was hidden from sight before the upper part was underwater. This happened at St. Paul's-wharf, Blackfriars-bridge, Temple-wharf, Southwark-bridge, and Hungerford; and I have no doubt would have occurred further up and down the river. Near the bridges the feculence rolled up in clouds so dense that they were visible at the surface, even in water of this kind.

The smell was very bad and common to the whole of the water; it was the same as that which now comes up from the gully holes in the streets; the whole river was for the time a real sewer. Having just returned from out of the country air, I was, perhaps, more affected by it than others; but I do not think I could have gone on to Lambeth or Chelsea, and I was glad to enter the streets for an atmosphere which, except near the sinkholes, I have found much sweeter than that on the river. I have thought it a duty to record these facts that they may be brought to the attention of those who exercise power or have a responsibility in relation to the condition of our river; there is nothing figurative in the words I have employed, or any approach to exaggeration; they are the simple truth. If there be sufficient authority to remove a putrescent pond from the neighbourhood of a few simple dwellings, surely the river which flows for so many miles through London ought not to be allowed to become a fermenting sewer. The condition in which I saw the Thames may perhaps be considered as exceptional, but it ought to be an impossible state, instead of which I fear it is rapidly becoming the general condition. If we neglect this subject, we cannot expect to do so with impunity; nor ought we to be surprised if, ere many years are over, a hot season give us sad proof of the folly of our carelessness.

I am, Sir, your obedient servant,
Royal Institution, July 7.

M. FARADAY.

Diabolic and Erotic

REAL MOLECULES WITH SILLY NAMES

In 2008, Professor Paul May, a chemist from the University of Bristol, wrote *Molecules with Silly or Unusual Names*, an expansion of the website on the same subject he had been maintaining since 1997. I strongly recommend you check it out, particularly if your sense of humour sometimes tends towards the puerile, as I'm afraid mine does.

DIUREA

Perhaps you won't be surprised to hear this is used in the fertiliser industry.

BASTARDANE

Similar to adamantane, this molecule contains a structural deviation from standard hydrocarbon caged arrangements and so became known as bastardane – the unwanted child.

PROFILACTIN

The unrecognised name of the complex formed by the binding of the proteins profilin and actin, with the latter being relevant to muscle contraction, appropriately enough.

PENGUINONE

So called because its two-dimensional structure resembles a penguin (sort of). It's a ketone, hence the –one ending.

URANATE

This is the name of the uranium oxide anions (negatively charged ions) such as UO_4^{2-}. Uranium nitrate is also known as uranyl nitrate. Maybe it's free during the day (the old ones are the best, they say).

ARSOLE

A ring molecule containing arsenic, which came to prominence thanks to the *Journal of Organometallic Chemistry* paper 'Studies on the Chemistry of Arsoles'. Does it help to say it was originally written in German?

EROTIC ACID

Not something to get you in the mood, this is really orotic acid. But Sigmund Freud might have been right about his slips and possibly should have studied more chemists, as the acid has been misspelt so many times that the name 'erotic acid' has now stuck.

ADAMANTANE

Adam Ant was one of the most recognisable pop frontmen of the early 1980s and co-writer of such classics as 'Prince Charming' and 'Stand and Deliver'. Unfortunately, the name of this chemical building block of diamond is unconnected, coming as it does from the Greek for indestructible, *adamas*.

APOLLAN-11-OL

This was first synthesised at the time of the Apollo 11 Moon landing. When drawn in two dimensions, it looks like a child's drawing of a rocket, complete with fins and exhaust, and has an alcohol group attached to carbon 11 that provides the -11-ol suffix. It's certainly a more memorable name than the molecule it's based on: 1,4,4,7- tetramethyltricyclo[5.3.1.0$^{2.6}$] undecane.

WINDOWPANE

A hydrocarbon with the formula C_9H_{12}, the structure of which is four squares joined together to make a bigger square, hence resembling a window. It has never been synthesised, but a version with one corner carbon missing – so that there were three squares joined together in an L-shape – has been, and was appropriately named 'broken window'.

The Music of the Spheres

*Mensus eram coelos,
nunc terrae metior umbras: Mens coelestis
erat, corporis umbra jacet.*

*[I used to measure the heavens,
now I shall measure the shadows of the earth.
Although my soul was from heaven, the
shadow of my body lies here.]*

**THE EPITAPH OF ASTRONOMER
JOHANNES KEPLER,
POSSIBLY SELF-ORIGINATED,
AS RECORDED BY HIS SON-IN-LAW**

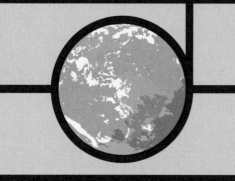

KEPLER LAYS DOWN THE LAW

As we've seen elsewhere in this book, German astronomer Johannes Kepler is a giant among giants in the history of science – an early believer in the Copernican system, in the early 1600s he would ultimately go beyond it to discover the true arrangement of the solar system. This was in part thanks to 16th-century Danish astronomer Tycho Brahe, a brilliant astronomical observer whose own desire, ironically, was to prove a system in which the Earth most definitely didn't move.

Using Brahe's extremely accurate data, Kepler was able to determine that Mars travelled along an elliptical orbit, as opposed to the circle proposed by all systems before, and that it moved with a regularly changing speed as it did so. To Kepler, a deeply religious man, this unexpected discovery posed a problem – why would God produce a system based on ellipses when circles seemed so much more harmonious? Kepler's answer, it turned out, came in the form of the ancient Pythagorean belief in the 'music of the spheres'– the idea that a celestial soundscape is produced in the heavens by the motion of astronomical bodies.

Kepler studied the speeds of each planet (there were six known planets at the time) at their aphelion (the point when a planet is farthest from the Sun and therefore at its slowest) and perihelion (the point when a planet is closest to the Sun and at its fastest). Incredibly, he worked out that the ratios of these pretty much matched those found in music. For example, the ratio of the speed of Mars at perihelion and Earth at aphelion was the same as that between two tones in a perfect fifth interval in music. This work was to prove of tangible benefit to science – a crucial consequence of Kepler's study of planet speeds was his third law.

Here are the three laws:

1

A planet's orbit is elliptical with the Sun at one focus (an ellipsis has two foci).

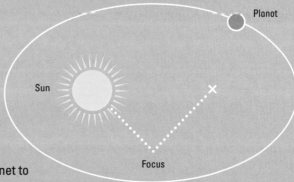

Planet

Sun

Focus

2

A line connecting a planet to the Sun sweeps out equal areas in equal periods of time – when the planet is closest to the Sun (the perihelion) and when it's furthest away (the aphelion).

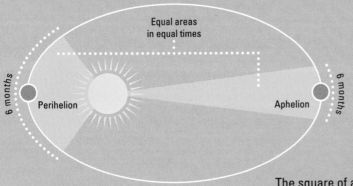

Equal areas in equal times

6 months

Perihelion

Aphelion

6 months

3

The square of a planet's period of revolution (T) is directly proportional to the cube of its mean distance from the Sun (a) (or $T^2 \propto a^3$).

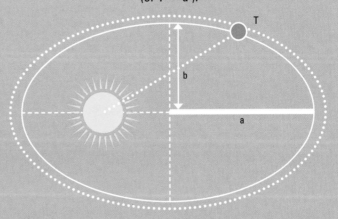

T

b

a

Unsung Heroine of Science

BRITAIN'S FIRST PROFESSIONAL FEMALE SCIENTIST

In 1762, William Herschel left his home town of Hanover, in north-west Germany, to try to make his way as a musician. He travelled to England, and after four rather difficult years in London he moved to Bath, where he had some success, taking on pupils and performing in the famous Pump Room. Financially on a better footing than he had ever been, William began to bring his siblings over to Bath, with his virtually uneducated sister Caroline arriving in 1772 to help run the household. William gave Caroline singing lessons, and soon she was performing in public on a regular basis, taking the lead soprano role in works such as the oratorio *Messiah* by George Frideric Handel.

Caroline Herschel

Ever since he had arrived in England, William's interest in astronomy had been growing and, after his sister's arrival, it began to compete with music for his time. He built his own telescope, turned his basement into a foundry to produce better mirrors than he could otherwise source and published papers on his astronomical observations. Caroline started to help her brother more and more in his studies, spending hours polishing his telescopic mirrors, to the detriment of her musical career.

All this hard work paid off when William discovered the planet that would later become known as Uranus (William, in a bout of sycophancy, named it Georgium Sidus, the Star of George, after King George III, but the competing name of Uranus was suggested by the German astronomer Johann Elert Bode, and this eventually won out). William's international fame was secured, and the following year the king offered him the position of royal astronomer at Windsor, with a pension of £200, which he accepted. Although this effectively ended Caroline's musical career, William now trained his sister to be an assistant astronomer.

AN OBSERVATORY AT HOME
William Herschel's (1738–1822) 6m
(20ft) reflector telescope, which he first
used two years after he discovered
Uranus in 1781.

Caroline was a meticulous observer, and she began to specialise in nebulae and comets. On 1 August 1786, she found her first comet and went on to discover seven more. In 1787, Caroline was awarded an annual pension of £50 from the king, which made her Britain's first professional female scientist. Her fame grew, and in 1790 she received a letter addressed simply 'Mlle Caroline Herschel, astronome célèbre, Slough' from the director of the Paris Observatory. Her scrupulous work continued, even after her brother married and she was forced to take her own lodgings. In 1828, she was awarded the Gold Medal of the Royal Astronomical Society of London. In later life, she moved back to Hanover and wrote her memoirs. She died in 1848, aged 97.

The following is a poem written by the Swedish-American writer and artist Siv Cedering, which in 1985 won her the Rhysling Award, founded by the Science Fiction Poetry Association. It's a wonderful modern interpretation of Caroline Herschel's life and a poignant reminder of some of the unsung heroines of science.

Letter from Caroline Herschel (1750–1848)

William is away, and I am minding
the heavens. I have discovered
eight new comets and three nebulae
never before seen by man,
and I am preparing an index to
Flamsteed's observations, together with
a catalogue of 560 stars omitted from
the British Catalogue, plus a list of errata
in that publication. William says
I have a way with numbers, so I handle
all the necessary reductions and
calculations. I also plan
every night's observation
schedule, for he says my intuition
helps me turn the telescope to discover
star cluster after star cluster.

I have helped him polish the mirrors
and lenses of our new telescope. It is
the largest in existence. Can you imagine
the thrill of turning it to some new
corner of the heavens to see
something never before seen
from Earth? I actually like
that he is busy with the Royal Society
and his club, for when I finish my other work

I can spend all night sweeping
the heavens.

Sometimes when I am alone
in the dark, and the universe reveals
yet another secret, I say the names
of my long lost sisters, forgotten
in the books that record our science.

Aganice of Thessaly,
Hyptia,
Hildegard,
Catherina Hevelius,
Maria Agnesi

– as if the stars themselves could remember.

Did you know that Hildegard
proposed a heliocentric universe
300 years before Copernicus? That she
wrote of universal gravitation 500 years
before Newton? But who would listen
to her? She was just a nun, a woman.
What is our age, if that age was dark?
As for my name, it will also be
forgotten, but I am not accused
of being a sorceress, like Aganice,
and the Christians do not threaten to
drag me to church, to murder me, like they did
Hyptia of Alexandria, the eloquent young
woman who devised the instruments
used to accurately measure the position
and motion of heavenly bodies.

However long we live, life is short, so I
work. And however important man becomes,
he is nothing compared to the stars.
There are secrets, dear sister, and it is
for us to reveal them. Your name, like mine,
is a song.

Write soon,
Caroline

*Coeloum
Perrupit Claustra
[He broke through the
barriers of the heavens]*

**EPITAPH TO WILLIAM HERSCHEL,
ST LAURENCE CHURCH,
UPTON-CUM-CHALVEY,
BERKSHIRE, ENGLAND**

Working with Black Bodies

THE BIRTH OF THE QUANTUM

In 1899, the American physicist Albert Michelson – who would win the 1907 Nobel Prize in Physics for his work in measuring the speed of light – made the sort of statement no scientist should ever be caught making. He declared 'the more important fundamental laws and facts of physical science have all been discovered, and these are now so firmly established that the possibility of their ever being supplanted in consequence of new discoveries is exceedingly remote'. This is the scientific equivalent of calling the *Titanic* 'unsinkable'.

Forty years earlier, the German physicist Gustav Kirchhoff had suggested the concept of the 'black body' as a theoretical ideal absorber and radiator of energy designed to investigate the connection between the temperature of an object and the colour it radiated. For example, iron will always glow red at the same temperature and will change to orange, yellow and then white as it gets hotter, again at consistent temperatures. The black body provided the perfect way to measure this connection. Being black, it's able to absorb all electromagnetic radiation (all the different forms of light) that falls on it. The body, in turn, emits radiation, which can then be measured. The hope was of establishing a formula that described how the intensity of radiation varied with wavelength at a given temperature (when iron glows red, the radiation emitted is most intense in the red part of the visible region of light).

One common problem for science is that it's often difficult to make ideal models into a reality, and it wasn't until the 1890s that creating a black body became a reasonable possibility. In 1893, German physicist Wilhelm Wien derived a formula that became known as 'Wien's displacement law' and which matched reasonably well with the experimental data being published at that time. The observed discrepancies – and there were some significant ones at the lower frequencies, the infrared end of the spectrum – were assumed by many to be the fault of the experiments rather than the law. Max Planck, professor of theoretical physics at Berlin University, wanted to establish a proof of the law and set about working on this. While he did, the debate regarding differences between the law's predictions and actual data from experiments grew until significant cracks in the former began to show.

Using better-quality data, Planck derived a formula that corresponded with the data well and which he felt was an improvement of Wien's. The only problem was that it wasn't clear to anyone, especially Planck, what exactly the formula meant. There was still a lot more work to be done, forcing Planck to test nearly every scientific assumption he held.

> *Scientists have odious manners, except when you prop up their theory; then you can borrow money off them.*
>
> **MARK TWAIN (1835–1910), AMERICAN WRITER**

A STUNNING SURPRISE

After incorporating the ideas of Austrian physicist Ludwig Boltzmann, who in the 1870s had produced a statistical description of the second law of thermodynamics founded on probabilities, Planck began to feel he was on the right track. But he was about to discover something totally unexpected. After toiling away, he hit on a formula that appeared to match theory and data, which he first presented to the world on 14 December 1900. It was the surprisingly simple, but also earth-shattering:

$$E = h\upsilon$$

where E is energy, υ is frequency and h is a constant.

The stunning implication of this formula was that energy, like matter, was made up of very small, discrete units, which Planck called 'energy quanta'. So, energy increased and decreased in minute steps like walking up and down stairs. Up until that point, an increase and decrease in energy had been thought of as being smooth and continuous, like rolling up and down a perfect hill. The h of the formula became known as Planck's constant, and with a value of 6.626×10^{-34} J s (joule-second) it's one of the smallest quantities in physics.

The quantum (singular of quanta) had been unleashed on the world. It was built on by Albert Einstein in 1905 in his paper on the photoelectric effect and extended by Danish physicist Niels Bohr in 1913 in his work on the structure of the atom.

This activity would eventually develop further into quantum mechanics and quantum electrodynamics. Without these and other quantum breakthroughs, today we'd have no transistor, superconductor or laser to name but three examples. And I wouldn't be writing this book on a computer. Max Planck's Nobel Prize in Physics 1918 was justly deserved.

A Pluto-sized Problem

THE DEFINITION OF A PLANET

After the discovery of Pluto in 1930, our solar system was said to comprise nine planets, a fact that could be rather nattily remembered with the neat mnemonic:

My Very Easy Method Just Speeds Up Naming Planets

However, a mini-crisis came about during the 1990s and 2000s when some other astronomical objects similar in size to Pluto – which was by then understood to be a member of a diffuse collection of objects called the Kuiper belt – were discovered. This led to a fierce debate regarding what exactly constituted a planet.

To resolve the issue, the International Astronomical Union (IAU) produced the following definition:

A PLANET IS A CELESTIAL BODY THAT

A

is in orbit around the Sun,

B

has sufficient mass for its self-gravity to overcome rigid body forces, so that it assumes a hydrostatic equilibrium (nearly round) shape, and

C

has cleared the neighbourhood around its orbit.

Pluto fails to satisfy the final condition and was downgraded
as a result (it, and similar objects, are now known as
plutoids). So, our solar system today contains only eight
planets, and a new mnemonic was born:

My Very Educated Mother
Just Served Us Nachos

Leaving a Trace in Space

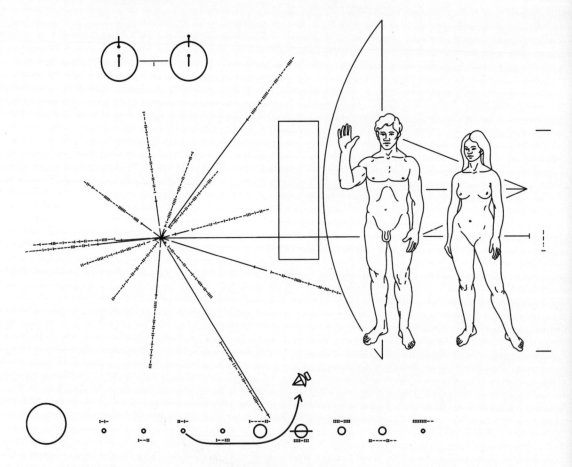

PIONEER'S POSTCARD FROM EARTH

Pioneer 10 was launched on 2 March 1972 and would become the first spacecraft to pass through the asteroid belt between Mars and Jupiter, and the first to get up close to Jupiter. It was also the first craft to leave the solar system and head into deep space. Because it was going further than anything humans had made before, attached to it was a plaque (shown above) created by the American science writer and astrophysicist Carl Sagan, along with his wife Linda Salzman Sagan – an artist and writer – and the astrophysicist Frank Drake. The plaque's purpose was to convey from who, when and where it had come to whoever might discover *Pioneer 10* during its journey.

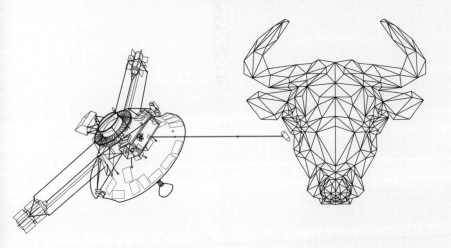

A week before the launch of *Pioneer 10*, NASA described the plaque's symbology as follows:

> *The Pioneer spacecraft, destined to be the first manmade object to escape from the solar system into interstellar space, carries this pictorial plaque. It is designed to show scientifically educated inhabitants of some other star system, who might intercept it millions of years from now, when Pioneer was launched, from where, and by what kind of beings. (With the hope that they would not invade Earth.) The design is etched into a 6 inch by 9 inch [15 by 23cm] gold-anodised aluminum plate, attached to the spacecraft's antenna support struts in a position to help shield it from erosion by interstellar dust. The radiating lines at left represent the positions of fourteen pulsars, a cosmic source of radio energy arranged to indicate our sun as the home star of our civilisation. The '1-' symbols at the ends of the lines are binary numbers that represent the frequencies of these pulsars at the time of launch of Pioneer relative to that of the hydrogen atom shown at the upper left with a '1' unity symbol. The hydrogen atom is thus used as a 'universal clock', and the regular decrease in the frequencies of the pulsars will enable another civilisation to determine the time that has elapsed since Pioneer was launched. The hydrogen is also used as a 'universal yardstick' for sizing the human figures and outline of the spacecraft shown on the right. The hydrogen wavelength, about 8 inches [20cm], multiplied by the binary number representing '8' shown next to the woman gives her height, 64 inches [163cm]. The figures represent the type of creature that created Pioneer. The man's hand is raised in a gesture of good will. Across the bottom are the planets, ranging outward from the Sun, with the spacecraft trajectory arching away from Earth, passing Mars, and swinging by Jupiter.*

Contact was lost with *Pioneer 10* in 2003 when it was more than 12 billion kilometres (7.5 billion miles) from Earth. The ship is now ghosting its way in the general direction of Aldebaran, a star more than 60 light-years away in the constellation Taurus. It will take the spacecraft more than 2 million years to reach it. A second plaque of the same design was attached to *Pioneer 11*, launched just over a year after *Pioneer 10*. This also journeyed to Jupiter before becoming the first spacecraft to explore Saturn and its rings up close. As with *Pioneer 10*, contact was lost as it headed into outer space, where it's now moving towards Sagittarius and the centre of the galaxy.

The True Measure of Things (Zooming Out)

6.5×10^1m
Wingspan of a
Boeing 747-400

3×10^2m
Height of the
Eiffel Tower

3.5×10^6m
Diameter of the
Moon

3.83×10^8m
Distance from the
Earth to the Moon

5.5×10^0m
Typical height of
a giraffe

4.3×10^5m
Distance of a
marathon

10^0m

10^3m

10^6m

10^9m

1.1×10^5m
1° of latitude at the equator

2×10^1m
Length of a
cricket pitch

1.3×10^7m
Diameter of
the Earth

8.85×10^3m
Height of Everest

1.4 x 10^{9}m
Diameter of the Sun

4 x 10^{16}m
Distance to Proxima Centauri, the
closest star to our solar system

1.5 x 10^{11}m
Distance from the Earth to the Sun (this is
the definition of 1 AU, or astronomical unit)

4.8 x 10^{12}m
Distance from the Sun to Pluto

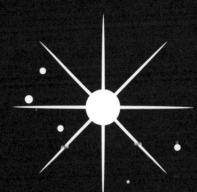

10^{12}m

10^{15}m

10^{18}m

10^{26}m

8.8 x 10^{26}m
Diameter of the observable universe

2.4 x 10^{12}m
Diameter of UY Scuti, the largest
known star

131

Nature and Nurture

EXTERNAL FACTORS

For a number of years, the general impression persisted that we were effectively slaves to our genes and that there wasn't much we could do about the hand nature had dealt us. Edward O. Wilson – American biologist and father of sociobiology (the study of evolution's influence on social behaviour) – famously paraphrased the words of English Victorian writer Samuel Butler ('A hen is only an egg's way of making another egg') when he said, 'The organism is only DNA's way of making more DNA'. Over recent years, this gene-centred view has, however, taken a bit of a battering, in part due to the rise of the study of epigenetics, which examines how genes can be differently expressed depending on other factors.

Meaning 'upon the gene', epigenetics looks at the mechanisms involved in gene expression – in essence, what switches genes on and off and also what adjusts how 'loud' or 'quiet' they are. It's now known that what your DNA 'says' isn't enough – it very much matters how it's then read by your cells. It turns out that external factors, such as the environment you live in and the food you eat, can have a large impact on this reading of your genetic code through the placing of markers on your DNA (although these don't affect its sequence).

In her 2011 book *The Epigenetics Revolution*, British biologist Nessa Carey makes the analogy that DNA is like a script. There have been thousands of performances of Shakespeare's *Hamlet* and, despite each using the same script, each has interpreted what Shakespeare wrote in a slightly different way. Taking the analogy one step further, the director and actors may have made notes on the script, which facilitated their interpretation by possibly emphasising, downplaying or even cutting a particular passage. This, in a sense, is the influence nurture has on your DNA and how it's read.

THE IMPORTANCE OF DIET

To give an extreme example of what this means, when a queen honeybee founds a new colony, there will be thousands of eggs. When these hatch, the resulting bee larvae feed on royal jelly given to them by what are known as nurse bees. After three days, this stops and the larvae are forced to make do with pollen and nectar, before eventually developing into worker bees, ready to start making honey. However, a very small number of larvae continue to be fed royal jelly and these eventually grow into queen bees. Incredibly, there is nothing that genetically distinguishes queens from workers and yet two very physically different bees, with different lifespans and physiology, will have developed simply because of their diet. What food they ate during their growth affected the expression of the genes in their DNA. This appears analogous to temperature-dependent sex determination in certain reptiles, such as the crocodile. Here, it's not possible to know what the sex of an offspring will be from its chromosomes. What affects the outcome is the temperature of incubation of the egg at key moments in its development.

Something similar, but perhaps even more remarkable, occurs in our own bodies – cell differentiation. In this process, stem cells, which, like the bees above, start with the same genetic code, somehow know how to become, say, specialised heart and kidney cells in their respective locations.

In his 1957 book, *The Strategy of the Genes*, the British developmental biologist Conrad Waddington proposed an epigenetic landscape to help to visualise this process.

In his metaphor, he describes a ball about to start a journey down a hill. This represents a cell with the *potential* to specialise. As it progresses, it can roll down any one of a series of valleys and troughs until it eventually ends at the bottom in one of many possible final destinations – these valleys represent different courses of development and the final destinations of the different possible cell types. By the time the cell reaches the bottom of the hill, it has specialised and become, for example, a heart, lung or kidney cell. To now get that cell to become a different type – to change a heart cell into a kidney cell, say – would be extremely hard because the ball would have to be pushed over some rather steep inclines. The best option would be to push the ball back along its original journey to restore its potential – if this were possible. Scientists are now beginning to understand how they might be able to do this.

Epigenetics has become one of science's most important areas of research and has huge implications for us all. We're starting to get a much better understanding of what happens at the level of the gene and how this affects our lives. You're likely to hear a lot more about it in the future.

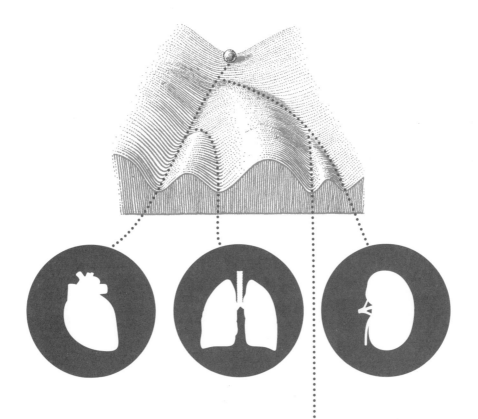

One Clear, Unchanged and Universal Light

CHASING THE MAGIC *c*

It wasn't until the 17th century that people began to determine the speed of light, designated by the unit *c*. Galileo describes an experiment to establish exactly that, in which lanterns a few kilometres apart are covered and then uncovered. This could never have worked because *c* is such a large value, but the sentiment was an important one.

The Danish astronomer Ole Rømer made the first significant measurement of the speed of light in 1676. By observing successive eclipses of the moons of Jupiter by Jupiter itself, he came to a value of 214,000km (133,000mi) per second. This is still quite some way off the true value but is at least in the right ballpark – that's to say, it's very, very fast. The next significant step in determining the speed of light was made by an Englishman James Bradley, who would go on to succeed Edmund Halley (after whom Halley's Comet is named) as Astronomer Royal in 1742. While attempting to detect stellar parallax (the difference in the direction a star lies in as seen from two separate points), he came up with the remarkably accurate 301,000km (193,000mi) per second.

French physicist Léon Foucault, American physicist Albert Michelson and others made further advances during the 19th century, and the 20th century saw various refinements until the current value of 299,792.458km (approximately 186,000mi) per second in a vacuum was adopted in 1983.

In 1905, Albert Einstein published his postulation that the speed of light was invariant; that it was the same whatever the motion of the observer. This is a key part of his special theory of relativity, which also includes the equation $E = mc^2$.

In September 2011, a team of scientists in Italy working on a project in collaboration with CERN announced results that appeared to show neutrino subatomic particles travelling faster than the speed of light. If true, this would have turned physics on its head and heralded the end of Einstein's theories of relativity. In November 2011, the experiment was repeated and gave the same result. In March 2012, however, results from a different group at the same Italian laboratory showed neutrinos travelling at exactly the speed of light, and it soon became clear there was a hardware problem with the first group's experiments – with a huge sigh of relief breathed by all.

Twinkle, twinkle little star, I don't wonder what you are, For by spectroscopic ken I know that you are hydrogen.

ANONYMOUS

Another Hue unto the Rainbow

AN INVISIBLE RAY

We often hear of people suffering for their art, but what about for science? Isaac Newton certainly suffered. During his investigations into colour, he often experimented on his own eyes. This included staring at the sun reflected in a mirror. It took him four days to recover his sight to an acceptable level, and he had recurring problems over the following months.

Even more shockingly, Newton also physically interfered with his own sight. In one of his notebooks, he talks of inserting a bodkin (a thick needle that was also sometimes used as a hair pin) between his eye and its socket as near to the back of the eye as possible. He would then press, to change the retina's curvature, resulting in his seeing 'white darke & coloured circles' as he continued to vary the pressure and movement. On a much less dangerous level, Newton carried out optical investigations using prisms he had bought at a local fair, with which he was able to show that white light was made up of the colours of the rainbow. In 1800, more than a century later, the British-German astronomer William Herschel began to investigate whether these colours corresponded to different temperatures. Using part of a chandelier as his prism, Herschel split sunlight into its spectral colours. He then placed two 'control' thermometers on either side of the spectrum before using a third to measure the temperature of each colour. He soon discovered that as he went from violet to green to red, the temperature increased. This was an exciting finding, but there was more to come. During Herschel's experiments, the sun had continued travelling on its course, which meant the split light had moved slightly as well. The thermometer had been measuring the red rays but was now just past them, yet Herschel was startled to find that the temperature measured was even higher – there appeared to be an invisible ray contained within sunlight. Via a well-constructed experiment, Herschel had accidentally discovered infrared radiation, a phenomenon that had been hypothesised in 1737 by French mathematician, physicist and noblewoman Émilie du Châtelet. Like waiting for buses, the discovery of ultraviolet radiation quickly followed a year later, in 1801.

$$v = \frac{c}{\lambda}$$

$$v = \frac{E}{h}$$

The four famous equations of Scottish scientist James Clerk Maxwell subsequently explained that visible light, infrared and ultraviolet radiation were all electromagnetic waves travelling at the speed of light but with differing frequencies and wavelengths. The equations also predicted many other forms of electromagnetic radiation, all related by the equation, left, where v is the frequency, c is the speed of light and λ is wavelength. Over the course of the following fifty years, the discoveries of microwaves, radio waves, X-rays and gamma rays completed the picture, resulting in the electromagnetic spectrum, bottom left.

The advent of quantum theory, which now argued that energy could only come in chunks (this is effectively what a photon is – a packet of energy), produced a different version of the equation:

E refers to photon energy and h is Planck's constant – with the incredibly small value of 6.626×10^{-34} J s (joule–second) – something its originator, the German physicist Max Planck, originally felt was an interim mathematical fudge to explain the results of an experiment, but which became a key part of this new science. The equation above also tells us that the energy of an electromagnetic wave is related directly to v – its frequency. The higher the frequency, the higher the energy. Radio waves therefore possess the least energy, whereas gamma rays have the highest. That's why gamma rays are so particularly harmful to human tissue, for example. (You may notice this equation doesn't tally with Herschel's findings regarding his thermometer's temperature increase as the frequency decreased – other factors were also involved.)

AIDE-MÉMOIRE
The electromagnetic spectrum from the lowest to highest energy. This order can be remembered using the mnemonic Rabbits Mate In Very Unusual eXpensive Gardens.

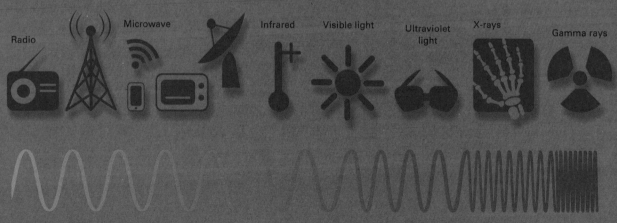

Radio · Microwave · Infrared · Visible light · Ultraviolet light · X-rays · Gamma rays

Decreasing wavelength, increasing frequency, increasing energy

On a New Kind of Ray

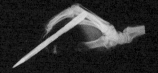

CELEBRATING RÖNTGEN'S DISCOVERY OF THE X-RAY

*If the hand be held between the discharge-tube and
the screen, the darker shadow of the bones is seen
within the slightly dark shadow-image of the hand
itself … For brevity's sake I shall use the expression
"rays"; and to distinguish them from others of this
name I shall call them "X-rays".*

So wrote the German physicist Wilhelm
Röntgen, winner of the first Nobel Prize in
Physics in 1901, when he publicly announced the
discovery he had made on 8 November 1895.

Of huge scientific importance, not least due to the separate and
equally significant transformation their discovery would produce
within medicine and physics, Röntgen's 'rays' caused a public
sensation right across Europe. By the beginning of February 1896,
there was already a thriving trade in 'shadow photographs'. *Punch*
magazine, ever one to chime with the public mood, published the
following witty poem barely a month after Röntgen had submitted
his preliminary communication to the Würzburg Physico-Medical
Society on 28 December 1895, and just two days after the journal
Nature had printed a translation of it.

*O, Röntgen, then the news is true,
And not a trick of idle rumour,
That bids us each beware of you,
And of your grim and graveyard humour.*

*We do not want, like Dr. Swift,
To take our flesh off and to pose in
Our bones, or show each little rift
And joint for you to poke your nose in.*

*We only crave to contemplate
Each other's usual full-dress photo;
Your worse than 'altogether' state
Of portraiture we bar in toto!*

*The fondest swain would scarcely prize
A picture of his lady's framework;
To gaze on this with yearning eyes
Would probably be voted tame work!*

*No, keep them for your epitaph,
These tombstone-souvenirs unpleasant;
Or go away and photograph
Mahatmas, spooks, and Mrs. B–s–nt!*

Centenary Icons of Science and Technology

* The Pilot ACE (Automatic Computing Engine) was one of the first computers to be built in the UK and was precursor to the more ambitious ACE designed by Alan Turing.

10. Electric telegraph

1. X-ray machine

2. Penicillin

9. Model T Ford

A SCIENCE MUSEUM TOP 10

To celebrate its centenary in 2009, the Science Museum in London selected ten 'centenary icons' of science and technology from right across its collection and invited people to vote for the one they felt was the greatest. The top place went to X-ray machines, which took 20 per cent of the total vote. The following two positions were also taken by discoveries related to medicine.

3. DNA double helix

8. Steam engine

4. Apollo 10 capsule

7. Pilot ACE computer

6. Stephenson's Rocket

5. V2 Rocket engine

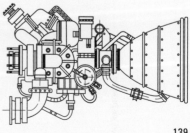

Chocolate, a Microwave and a Ruler

MEASURING THE SPEED OF LIGHT

It took more than 300 years of experimentation and refinement to arrive at the figure for the speed of light that we use as standard today. There's also a method for determining the speed yourself that might seem more than a little surprising. All you need is some kind of food that can melt (chocolate is good but you can also use marshmallows or cheese), a microwave oven, a microwave-safe dish to put the food in and a ruler.

1   Place the food on the dish.

2   Remove the turntable from the microwave – it's important that the dish can't move.

3   Put the dish in the microwave.

4   Cook on a low heat until it's clear the food is beginning to melt in spots. Begin by trying 30 seconds. These spots relate to the peaks of the 'wave' – the distance between two peaks is half a wavelength.

5   Once the melted spots appear, remove the dish and measure the distance between the centres of these spots. One distance should repeat again and again.

Now look on the microwave casing (probably on the back) to find its frequency – this is typically 2.45GHz.

We know that $c = \lambda v$, or the speed of light = wavelength multiplied by frequency.

So, v is the frequency of the microwave. If it's 2.45GHz, then the figure you'll use in your calculation will be 2,450,000,000 (whatever the frequency listed is, it will almost certainly be in gigahertz – 1GHz is 1,000,000,000, so make sure your calculation reflects that). You now need to multiply this by λ, which is double the distance you measured in metres (for example, 15cm is 0.15m, which doubled becomes 0.30m). See how close to the true speed of light – 299,792,458m per second – you get.

HOT CHOCOLATE

The melted spots in the chocolate are a result of the crests and troughs of the standing waves the oven produces. The distance between a crest and its adjacent trough is half the wavelength.

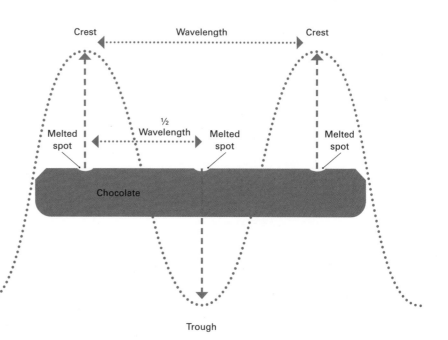

141

'Like a swan that can't settle'

A LITERARY GIANT'S VIEW OF RELATIVITY

On 15 June 1921, the English author D. H. Lawrence wrote to a friend to confirm receipt of the copy he had sent of Albert Einstein's *Relativity: The Special and General Theory*. The book was written with the non-specialist in mind, and Lawrence devoured its contents in a single day.

A year later, he applauded the new theory, declaring himself to be 'very pleased with Mr Einstein for knocking that eternal axis out of the universe. The universe isn't a spinning wheel. It is a cloud of bees flying and veering round'. Lawrence then shifted his attention from the universe to human relationships, proclaiming in his *Fantasia of the Unconscious* that 'we are in sad need of a theory of human relativity'.

Ironically, Lawrence would go on to admit to not being able to grasp the new relativity, along with quantum theory, in his poem 'Relativity', included in his collection *Pansies*, which was published in 1929, a year before his death.

D. H. Lawrence

'RELATIVITY'

I like relativity and quantum theories
because I don't understand them
and they make me feel as if space
shifted about like a swan that
*   can't settle,*
refusing to sit still and be measured;
and as if the atom were an
*   impulsive thing*
always changing its mind.

It's All Relative

EINSTEIN –
GREATEST OF ALL TIME?

If we were ever compelled to answer the thorny question of who the greatest scientist ever to have lived was, it would be extremely hard to look at anyone other than Isaac Newton and Albert Einstein as finalists.

Einstein's *annus mirabilis* was 1905. While still a patent clerk in Bern, Switzerland, he published five papers. One was on Brownian motion, and effectively put an end to the argument as to whether atoms truly existed or not; another was on the discovery of the photoelectric effect, for which he won the Nobel Prize in Physics 1921; and perhaps most importantly, one paper was on his special theory of relativity, and included the line 'if a body gives off the energy L in the form of radiation, its mass diminishes by E/c^2'. This led to probably the most famous equation of all time, $E = mc^2$.

In 1916, Einstein expanded the special relativity paper, so that it could take gravity and its effects into account, heralding the birth of the general theory of relativity. It was the apparent proof of this theory through the observation of a solar eclipse in 1919 that catapulted Einstein into the public consciousness. The following news report from *The New York Times* from the end of that year is full of charm and provides a rare glimpse into the early attempts to explain Einstein and his theory to the world at large.

In science, the credit goes to the man who convinces the world, not to the man to whom the idea first occurs.

FRANCIS DARWIN (1848–1925), ENGLISH BOTANIST AND SON OF CHARLES DARWIN

EINSTEIN EXPOUNDS HIS NEW THEORY

It Discards Absolute Time and Space,
Recognizing Them Only as Related to Moving Systems

IMPROVES ON NEWTON

Whose Approximations Hold for Most Motions,
but Not Those of the Highest Velocity

INSPIRED AS NEWTON WAS

But by the Fall of Man from a Roof Instead of the Fall of an Apple

Special Cable to *The New York Times*

BERLIN, Dec. 2. – Now that the Royal Society, at its meeting in London on Nov. 6, has put the stamp of its official authority on Dr. Albert Einstein's much-debated new 'theory of relativity', man's conception of the universe seems likely to undergo radical changes. Indeed, there are German savants who believe that since the promulgation of Newton's theory of gravitation no discovery of such importance has been made in the world of science.

When *The New York Times* correspondent called at his home to gather from his own lips an interpretation of what to laymen must appear the book with the seven seals, Dr. Einstein himself modestly put aside the suggestion that his theory might have the same revolutionary effect on the human mind as Newton's theses. The doctor lives on the top floor of a fashionable apartment house on one of the few elevated spots in Berlin – so to say, close to the stars which he studies, not with a telescope, but rather with the mental eye, and so far only as they come within the range of his mathematical formulae; for he is not an astronomer but a physicist.

It was from his lofty library, in which this conversation took place, that he observed years ago a man dropping from a neighboring roof – luckily on a pile of soft rubbish – and escaping almost without injury. This man told Dr. Einstein that in falling he experienced no sensation commonly considered as the effect of gravity, which, according to Newton's theory, would pull him down violently toward the Earth. This incident, followed by further researches along the same line, started in his mind a complicated chain of thoughts leading finally, as he expressed it, 'not to a disavowal of Newton's theory of gravitation, but to a sublimation or supplement of it.'

When he read in the message from *The Times* requesting the interview a reference to Dr. Einstein's statement to his publishers on the submission of his last book that not more than twelve persons in all the world could understand it, coupled with the editor's request that Dr. Einstein put his theory in terms comprehensible to a larger number than twelve, the doctor laughed good-naturedly, but still insisted on the difficulty of making himself understood by laymen.

'However,' he said, 'I am trying to talk as plainly as possible. To begin with the difference between my conception and Newton's law of gravitation: please imagine the Earth removed, and in its place suspended a box as big as a room or a whole house and inside a man naturally floating in the centre, there being no force whatever pulling him. Imagine, further, this box being, by a rope or other contrivance, suddenly jerked to one side, which is scientifically termed "difform motion," as opposed to "uniform motion." The person would then naturally reach bottom on the opposite side. The result would consequently be the same as if he obeyed Newton's law of gravitation, while, in fact, there is no gravitation exerted whatever, which proves that difform motion will in every case produce the same effects as gravitation.

'I have applied this new idea to every kind of difform motion and have thus developed mathematical formulas which I am convinced give more precise results than those based on Newton's theory. Newton's formulas, however, are such close approximations that it was difficult to find by observation any obvious disagreement with experience.

'One such case, however, was presented by the motion of the planet Mercury, which for a long time baffled astronomers. This is now completely cleared up by my formulas, as the Astronomer Royal, Sir Frank Dyson, stated at the meeting of the Royal Society.

'Another case was the deflection of rays of light when passing through the field of gravitation. No such deflections are explicable by Newton's theory of gravitation.

Albert Einstein

'According to my theory of difform motion, such deflections must take place when rays pass close to any gravitating mass, difform motion then coming into activity.

'The crucial test was supplied by the last total solar eclipse, when observations proved that the rays of fixed stars, having to pass close to the Sun to reach the Earth, were deflected the exact amount demanded by my formulas, confirming my idea that what so far has been regarded as the effect of gravitation is really the effect of difform motion. Elaborate apparatus and the closest and the most indefatigable attention to the difficult task enabled that English expedition, composed of the most talented scientists, to reach those conclusions.'

'Why is your idea termed "the theory of relativity?"' asked the correspondent.

'The term relativity refers to time and space,' Dr. Einstein replied. 'According to Galileo and Newton, time and space were absolute entities, and the moving systems of the universe were dependent on this absolute time and space. On this conception was built the science of mechanics. The resulting formulas sufficed for all motions of a slow nature; it was found, however, that they would not conform to the rapid motions apparent in electrodynamics.

'This led the Dutch professor, Lorenz, and myself to develop the theory of special relativity. Briefly, it discards absolute time and space and makes them in every instance relative to moving systems. By this theory all phenomena in electrodynamics, as well as mechanics, hitherto irreducible by the old formulae – and there are multitudes – were satisfactorily explained.

'Till now it was believed that time and space existed by themselves, even if there was nothing else – no Sun, no Earth, no stars – while now we know that time and space are not the vessel for the universe, but could not exist at all if there were no contents, namely, no Sun, Earth, and other celestial bodies.

'This special relativity, forming the first part of my theory, relates to all systems moving with uniform motion; that is, moving in a straight line with equal velocity.

'Gradually I was led to the idea, seeming a very paradox in science, that it might apply equally to all moving systems, even of difform motions, and thus I developed the conception of general relativity which forms the second part of my theory.

'It was during the development of the formulas for difform motions that the incident of the man falling from the roof gave me the idea that gravitation might be explained by difform motion.'

'If there is no absolute time or space, supposedly forming the vessel of the universe,' the correspondent asked, 'what becomes of the ether?'

'There is no ether, as hitherto conceived by science, which is proved by the well-known experiment of the celebrated American savant, Michelson, showing that no influence by the motion of the Earth on the ether is perceptible through change in velocity of light, such as ought to be produced if the old conception were true.'

'Are you yourself absolutely convinced of the correctness of this revolutionary theory of relativity, or are there still any reservations?'

'Yes, I am,' Dr. Einstein answered. 'My theory is confirmed by the two crucial cases mentioned before. But there is still one test outstanding, namely the spectroscopic. According to my theory, the lines of the spectra of fixed stars must be slightly shifted through the influence of gravitation exerted by the very stars from which they emanate. So far, however, the results of the examinations have been contradictory; but I have no doubt of final confirmation even through this test*.'

Just then an old grandfather clock in the library chimed the mid-day hour, reminding Dr. Einstein of some appointment in another part of Berlin, and old-fashioned time and space enforced their wonted absolute tyranny over him who had spoken so contemptuously of their existence, this terminating the interview.

*Later experiments would show Einstein had been right about this also.

A Popular Choice

ROYAL SOCIETY PRIZE FOR SCIENCE BOOKS

First awarded in 1988, the Royal Society science books prize was established 'with the aim of encouraging the writing, publishing and reading of good and accessible popular science books'. It has had various sponsors during that time and has been variously referred to as the 'Rhône-Poulenc Prize for Science Books' (1990–2000), the 'Aventis Prize for Science Books' (2001–2006), the 'Royal Society Prize for Science Books' (2007–2010), the 'Royal Society Winton Prize for Science Books' (2011–2015) and the 'Royal Society Insight Investment Science Book Prize' (2016 to date). Except for the years 2009–2010, a companion junior book prize has run concurrently.

1999 *The Man Who Loved Only Numbers*, Paul Hoffman

1998 *Guns, Germs and Steel*, Jared Diamond

1997 *The Wisdom of Bones*, Alan Walker and Pat Shipman

1996 *Plague's Progress*, Arno Karlen

1995 *The Consumer's Good Chemical Guide*, John Emsley

1991 *Wonderful Life*, Stephen Jay Gould

1994 *The Language of Genes*, Steve Jones

1990 *The Emperor's New Mind*, Roger Penrose

1993 *The Making of Memory*, Steven Rose

1989 *Bones of Contention*, Robert Lewin

1988 *Living with Risk*, British Medical Association

1992 *The Rise and Fall of the Third Chimpanzee*, Jared Diamond

2021 *Entangled Life*, Merlin Sheldrake

2020 *Explaining Humans*, Camilla Pang

2019 *Invisible Women*, Caroline Criado Perez

2018 *Inventing Ourselves*, Sarah-Jayne Blakemore

2017 *Testosterone Rex*, Cordelia Fine

2016 *The Invention of Nature*, Andrea Wulf

2015 *Adventures in the Anthropocene*, Gaia Vince

2014 *Stuff Matters*, Mark Miodownik

2013 *The Particle at the End of the Universe*, Sean Carroll

2012 *The Information*, James Gleick

2011 *The Wavewatcher's Companion*, Gavin Pretor-Pinney

2010 *Life Ascending*, Nick Lane

2009 *The Age of Wonder*, Richard Holmes

2008 *Six Degrees*, Mark Lynas

2007 *Stumbling on Happiness*, Daniel Gilbert

2006 *Electric Universe*, David Bodanis

2005 *Critical Mass*, Philip Ball

2004 *A Short History of Nearly Everything*, Bill Bryson

2003 *Right Hand, Left Hand*, Chris McManus

2002 *The Universe in a Nutshell*, Stephen Hawking

2001 *Mapping the Deep*, Robert Kunzig

2000 *The Elegant Universe*, Brian Greene

Tearing Up the Rules

THE BOOK THAT CAUSED A PARADIGM SHIFT

In 1962, a history and philosophy of science monograph was published, which would go on to sell well over 1 million copies and introduce the world to the concept of the paradigm shift. By looking at how science has been practised throughout history, *The Structure of Scientific Revolutions*, by American science philosopher Thomas Kuhn, seriously questioned the then standard view that science progresses by building incrementally on what has gone before and can therefore be said to be edging steadily towards a *true* description of the world.

Kuhn began by describing what he referred to as 'normal science'. He gave Newtonian mechanics as an example of 'normal science', consisting of a mature set of scientific ideas that subsequently determine what research the scientific community does, typically by looking for what they expect to find. However, over time, the number of unexplained experimental anomalies within this 'paradigm' grows and a crisis can occur. This, in turn, can lead to a revolution. In these circumstances, a collection of scientists typically continues to work within the current paradigm, attempting to rescue it. At the same time, a splinter group will probably help to give birth to an entirely new paradigm, which over time might gain the support of more and more scientists until, at some point, science undergoes a paradigm shift, whereby the world view that was initially revolutionary becomes the generally accepted status quo.

A classic example is Copernicus's proposal that the Earth revolved around the Sun. But this shift is rarely smooth. As Max Planck – who prompted one such revolution when his 'black body problem' resulted in the birth of quantum theory (see page 124) – put it, 'a new scientific truth does not triumph by convincing its opponents and making them see the light, but rather because its opponents eventually die, and a new generation grows up that is familiar with it'.

One consequence of Kuhn's book was that it seemed to suggest the notion of 'incommensurability' in science – the view that concepts discussed in a new paradigm bore no connection with those in the previous one. This notion implied, for example, that gravity in Einstein's general theory of relativity was *entirely* different to Newton's, and the two scientists were talking about two different things (there was an argument that the mathematics underpinning them reinforced this view). If that was the case, then science couldn't exactly be said to be building on what had come before.

Instead, the inference was that science just moved to a completely different building. Naturally, this interpretation didn't go down terribly well within the scientific community. Kuhn's ideas also contained many problems of their own, not least in only really describing the biggest moments in the history of science – such as Copernicus's proposal, quantum theory and relativity – nearly all of which were centred around the physical sciences. According to the Canadian philosopher Ian Hacking, when the life sciences are considered (and their role within science has increased considerably since Kuhn wrote his book), Kuhn's diagnosis is much less convincing because these tend to be much less theory-driven, and much more analytical, than the physical sciences. But there can be little doubt of the impact *The Structure of Scientific Revolutions* has had – it provoked its very own revolution in how the practice of science is understood.

First get your facts; and then distort them at your leisure.
MARK TWAIN

151

Left, Right, Left, Right

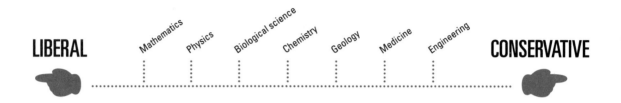

LIBERAL — Mathematics · Physics · Biological science · Chemistry · Geology · Medicine · Engineering — CONSERVATIVE

POLITICAL PERSUASIONS
Within the sciences, disciplines ranked as above in terms of most liberal to most conservative.

THE POLITICS OF SCIENTISTS

In 1969, just months after Richard Nixon was elected president of the United States of America, a questionnaire was sent to more than 60,000 American college and university professors that would provide researchers with more than 300 pieces of information on these individuals, including many relating to their political orientation. Among those who responded were 1,707 physicists, 1,884 chemists, 2,916 mathematicians, 812 geologists, 4,567 biological scientists, 2,395 faculty members in colleges of medicine and 4,382 engineers. Three years later, a paper detailing the general political differences between disciplines, and between high-achieving scientists and their more 'rank and file colleagues', was published in the journal *Science*. Some very interesting trends were observed.

Researchers discovered a clear correlation between particular disciplines and the political orientations of professors within them. So, social scientists and humanities professors were more liberal than scientists, who were more liberal than engineers.

The researchers felt their results showed that there was a relationship between how intellectual a discipline was and how liberal and change-orientated its practitioners were – the more practically based the discipline was, the more conservative its practitioners. Researchers also discovered that within disciplines, the higher achievers had more liberal political views, while the more 'rank and file' had more conservative views.

Evolution – It Runs in the Family

ERASMUS DARWIN'S SUBLIME IMAGES OF NATURE

Nearly sixty years before his grandson Charles's *On the Origin of Species*, physician Erasmus Darwin (1731–1802) wrote *The Temple of Nature, or, The Origin of Society*, a poetic hymn to evolution, which discussed the 'production of life', the 'reproduction of life', 'progress of the mind' and 'good and evil'.

Darwin was a member of the Lunar Society of Birmingham (depicted below, in an illustration from 1870), which also included the steam engine partners James Watt and Matthew Boulton, Joseph Priestley (the discoverer of oxygen) and the innovative potter Josiah Wedgwood. The physician's final work, *The Temple of Nature* sought 'simply to amuse by bringing distinctly to the imagination the beautiful and sublime images of the operations of Nature in the order [...] in which the progressive course of time presented them'.

From Canto I of
The Temple of Nature

Organic Life beneath the shoreless waves
Was born and nurs'd in Ocean's pearly caves;
First forms minute, unseen by spheric glass,
Move on the mud, or pierce the watery mass;
These, as successive generations bloom,
New powers acquire, and larger limbs assume;
Whence countless groups of vegetation spring,
And breathing realms of fin, and feet,
*   and wing.*
Thus the tall Oak, the giant of the wood,
Which bears Britannia's thunders on the flood;
The Whale, unmeasured monster of the main,
The lordly Lion, monarch of the plain,
The Eagle soaring in the realms of air,
Whose eye undazzled drinks the solar glare,
Imperious man, who rules the bestial crowd,
Of language, reason, and reflection proud,
With brow erect who scorns this earthy sod,
And styles himself the image of his God;
Arose from rudiments of form and sense,
An embryon point, or microscopic ens!

> *Success in research needs four Gs: Glück, Geduld, Geschick und Geld.*
> *[Luck, patience, skill and money]*
>
> **PAUL EHRLICH (1854–1915), GERMAN SCIENTIST, AS QUOTED IN M. PERUTZ, 'RITA AND THE FOUR GS', *NATURE*, 338, 791 (1988)**

There's Electricity in the Air

TYPES OF IONISING RADIATION

The type of radiation being referred to whenever radioactivity is mentioned is ionising radiation. It is, as the term rather heavily implies, radiation that can ionise – that is, produce electrically charged particles in anything it strikes, thereby changing the chemistry of that thing and making it more reactive. The effect of this on organisms can be significant.

We are exposed to ionising radiation every minute of every day, much of it in the form of background radiation. This includes cosmic rays, rocks in the ground, radon gas, water and food. Bananas, for example, contain naturally occurring potassium-40, a radioactive isotope of potassium. Incredibly, there's even something known as the 'banana-equivalent dose', which was created in an attempt to contextualise artificial radiation exposures, especially to the general public (an X-ray screening at a US airport is roughly two-and-a-half times a banana-equivalent dose).

> *If you're not part of the solution, you're part of the precipitate.*
> **ANONYMOUS**

Once general background radiation is accounted for, other sources include people's occupations (aeroplane crew, for example, are exposed to a greater amount than average of cosmic rays), medical procedures (including X-rays and CT scans), industrial uses and the legacy of past nuclear disasters and explosions.

Types of ionising radiation include:

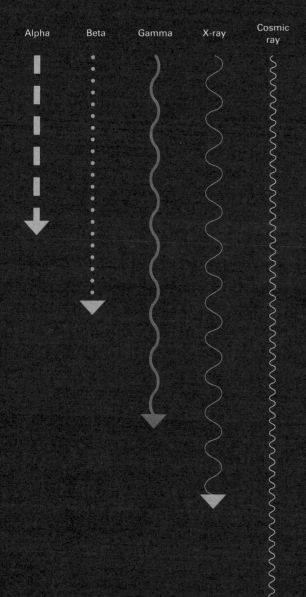

Alpha, Beta, Gamma, X-ray, Cosmic ray

Alpha (α) particles – Identical to the *nucleus* of a helium atom, alpha particles contain two protons and two neutrons bound together. They are by far the heaviest of the types of radiation and are easily stopped by a sheet of paper or the first layer of the skin. However, they can be extremely dangerous to humans if a source is ingested. In late 2006, the Russian journalist and ex-KGB officer Alexander Litvinenko was poisoned with polonium-210, a radioactive isotope of polonium, possibly administered to him in a cup of tea. It decays by alpha particle emission only.

Beta (β) particles – These are high-speed electrons and, as such, are also high-energy. They can penetrate further than alpha particles – through about 1–2cm (0.4–0.8in) of skin – but can be stopped by a few millimetres of aluminium.

Gamma (γ) rays – These are a high-energy type of electromagnetic radiation and therefore have no mass. They are extremely penetrative and can pass through the human body, causing great damage.

X-rays – Like gamma rays, although with a lower frequency, longer wavelength and less energy, X-rays are a type of electromagnetic radiation and can pass through the human body. They can cause harm in large doses, and so medical and dental use is tightly controlled.

Cosmic rays – In spite of their name, cosmic rays aren't actually rays but highly charged subatomic particles – typically protons but also electrons and atomic nuclei. Their origins lie in a variety of sources located in outer space, such as exploding supernovae. It's estimated that about a million cosmic rays pass through our bodies each night while we sleep. As the rays can penetrate far, particle physics experiments are performed deep underground to minimise any ray-induced interference with the controlled particle collisions being performed.

Checking the Dosage

NATURALLY OCCURRING RADIATION

Our world has always been radioactive. The following pie chart, adapted from one produced by the World Nuclear Association (WNA) – a body that 'represents the people and organisations of the global nuclear profession' – averages the sources of radiation we're exposed to every year. According to the WNA, 85 per cent of the radiation we're exposed to comes from natural sources (although this will naturally depend on a variety of factors, including where you live).

42% Radon

11% Food & drinking water

14% Medicine

1% Nuclear industry

18% Buildings and soil

Sources of radiation   14% Cosmic

When it comes to atoms, language can be used only as in poetry.

NIELS BOHR (1885–1962), DANISH PHYSICIST

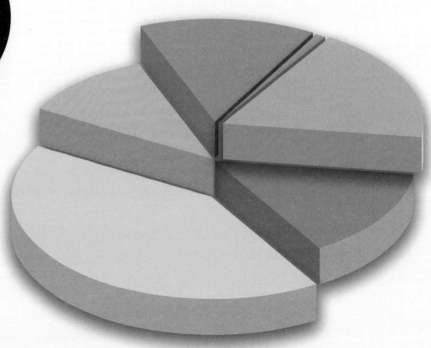

The general public has long been divided into two parts: those who think science can do anything, and those who are afraid it will.

DIXY LEE RAY (1914–1994), AMERICAN MARINE BIOLOGIST, FORMER CHAIRWOMAN OF THE ATOMIC ENERGY COMMISSION AND WASHINGTON STATE'S FIRST FEMALE GOVERNOR

There are various radiation units. These include the international system of units (SI)-derived becquerel (Bq), gray (Gy) and sievert (Sv). The becquerel is a measure of the activity of a radioactive source. Here, 1Bq equals 1 decay per second (1 decay equates to 1 emission of ionising radiation – for example, an α-particle). The gray is the unit that measures the absorbed dose of radiation and refers to the amount of energy absorbed as related to mass. So, 1 Gy equals 1 joule per kilogram of the mass of the thing doing the absorbing, such as the human body.

However, a measurement of an absorbed dose isn't necessarily the best way of assessing its biological effects, and therefore its danger. Different things, such as different human tissue or organs, are affected in different ways. As such, to best appreciate the risk from radiation it's necessary to translate the absorbed dose into the sievert, known as the unit of equivalent dose. This is done by multiplying the absorbed dose by a radiation weighting factor relevant to the type and energy of the radiation. The latter are determined, and regularly updated, by the International Commission on Radiological Protection (ICRP). So, the sievert is used when we want to judge the effect of radiation on us. It matters how big the dose is, the type of radiation and over what period it's given. The following table compares different dosages (pay attention to the time periods involved).

Dose, millisieverts (mSv)	Event
10,000	Given as a single dose, this would prove fatal within weeks
6,000	Typical dosage recorded in those Chernobyl workers who died within a month of the nuclear disaster in 1986
5,000	As a single dose, this would kill half of those exposed to it within a month
1,000	As a single dose, this would cause radiation sickness, including nausea and lower white blood cell count, although it would not prove fatal
1,000	As an accumulated dosage, this amount is estimated to cause a fatal cancer many years later in 5 per cent of people
400	Maximum radiation levels recorded at Japan's Fukushima nuclear plant, per hour, in 2011
350	Exposure of Chernobyl residents who were relocated after the blast in 1986
260	Average annual background level in some areas of Ramsar in Iran (no identified health effects)
100	Lowest annual dose at which an increase in occurrences of cancer is clearly evident in recipients
20	Current annual limit for employees in nuclear industry
16	CT scan: heart
10	Full-body CT scan
9	Airline crew flying New York to Tokyo polar route, annual exposure
5.9	Average annual background radiation per person in Finland
2.7	Average annual background radiation per person in the UK
0.2	Chest X-ray
0.005	135g (4.75oz) bag of Brazil nuts/dental X-ray

Gateway into Radioactivity

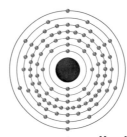

URANIUM'S FOURTEEN STEPS OF DECAY

Uranium – element number 92 in the periodic table and naturally occurring – is probably the most famous of radioactive substances. It was through experiments on uranium that the phenomenon of radioactivity was first recognised in 1896 by French physicist Henri Becquerel, who shared the Nobel Prize in Physics 1903 with fellow French scientists Marie and Pierre Curie. Uranium was also the radioactive substance involved in the first controlled nuclear chain reaction and lay at the heart of the atomic bomb dropped on the Japanese city of Hiroshima in 1945. As such, uranium provides the perfect gateway into understanding some of the characteristics of radioactive materials.

More than 99 per cent of all natural uranium comes in the form of its isotope uranium-238 (the isotopes of an element differ only in the number of neutrons in their nuclei – there are five other isotopes of uranium, one of which, uranium-234, is listed in the decay chain below). Uranium-238 is unstable and gradually decays. It has a half-life (the time it takes for a mass of it to decrease by half) of more than 4 billion years. Furthermore, uranium decays naturally through a series of fourteen steps, with radioactive emission occurring at each one, finally resulting in the isotope lead-206. Despite not resulting in the alchemist's dream of gold, this transmutation of one substance into another is one of nature's most fascinating processes.

Uranium-238
↓ → α, 4.5 billion years

Thorium-234
↓ → β, 24 days

Protactinium-234
↓ → β, 1.2 minutes

Uranium-234
↓ → α, 240,000 years

Thorium-230
↓ → α, 77,000 years

Radium-226
↓ → α, 1,600 years

Radon-222
↓ → α, 3.8 days

Polonium-218
↓ → α, 3.1 minutes

Lead-214
↓ → β, 27 minutes

Bismuth-214
↓ → β, 20 minutes

Polonium-214
↓ → α, 160 microseconds

Lead-210
↓ → β, 22 years

Bismuth-210
↓ → β, 5 days

Polonium-210
↓ → α, 140 days

Lead-206 (stable)

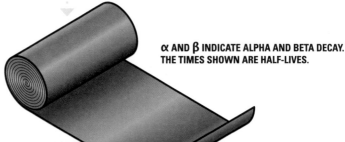

α AND β INDICATE ALPHA AND BETA DECAY. THE TIMES SHOWN ARE HALF-LIVES.

A Sports Hall Goes Nuclear

THE FIRST NUCLEAR REACTOR

It's a pretty incredible fact that the world's
first nuclear reactor was constructed in a
racquets court under the spectator stands of
the University of Chicago's – at that time
– recently closed football stadium, Stagg Field
(today's stadium is several blocks northwest
of where the original stood).

Built in the late autumn of 1942, Chicago Pile-1, as it was
known, consisted of layers of solid graphite blocks with other
layers containing uranium metal and/or uranium oxide fuel in
between them. Interspersed at regular intervals were
horizontal shafts in which three sets of 'control rods', made of
more graphite but also cadmium plated, could be removed or
reinserted. It was these rods that allowed the team of scientists
to manipulate the anticipated nuclear chain reaction. Leading
the project was the Italian physicist Enrico Fermi, who had
emigrated to the United States four years earlier after picking
up his Nobel Prize in Sweden.

By 2 December, 'the pile' was ready to be tested.
Beginning at 9.45am, over the course of the following six
hours the control rods were slowly and carefully withdrawn by
small degrees. At 3.36pm the reaction became self-sustaining
and was allowed to continue for a further 28 minutes. Then
Fermi instructed the main control rod to be reinserted, and the
reaction stopped. It was a remarkable achievement.

Later that day, Arthur Compton, an American physicist
and the man in overall charge of the project, phoned James B.
Conant, president of Harvard University, informing him 'the
Italian navigator ha[d] landed in the New World'. 'How were
the natives?' asked Conant. 'Very friendly,' was Compton's
reply. But the joy in the achievement wasn't unanimous. Once
most of those present had left for the day, one of Fermi's fellow
scientists, the Hungarian-American physicist Leó Szilárd, let
him know he thought the day would go down as a 'black day
in the history of mankind'.

PILE-1
A scale model of the Chicago Pile nuclear reactor.

An Unholy Trinity

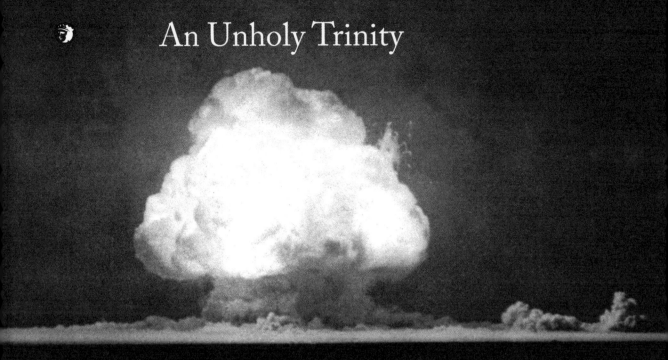

MAKING THE FIRST ATOMIC BOMB

Just over two and a half years after Enrico Fermi led a team of scientists to produce the first controlled nuclear chain reaction (see page 158), the first atomic bomb was detonated in the Jornada del Muerto desert, New Mexico, at 5.30am on 16 July 1945. Codenamed 'Trinity', it was the culmination of the gargantuan Manhattan Project, a scientific program costing more than $2 billion dollars (more than $30 billion today). The project was based primarily at Los Alamos, New Mexico.

As the book *Trinity Site* by the National Atomic Museum (now the National Museum of Nuclear Science and History) makes clear, the public wasn't officially made aware of the test until the atomic bomb had been dropped on Hiroshima, Japan, on 6 August. But many people were already aware that something extremely significant had taken place. The light from Trinity's blast could be seen across the entire state of New Mexico and beyond. Minutes later, the ensuing

mushroom cloud climbed to a height of more than 11,500m (38,000ft). The power of the detonation had been equal to 20,000 tonnes of TNT, or the payload of about 2,000 of the largest US Second World War bombers, the B-29 Superfortress. The bomb dropped on Hiroshima was approximately four-fifths as powerful. Trinity destroyed all living things within a mile.

William L. Laurence, a reporter for *The New York Times* and the Manhattan Project's official journalist, described the test with a terrible beauty when he wrote:

> *Just at that instant there rose as if from the bowels of the earth a light not of this world, the light of many suns in one. It was a sunrise such as the world had never seen, a great green supersun climbing in a fraction of a second to a height of more than eight thousand feet, rising ever higher until it touched the clouds, lighting up earth and sky all around with dazzling luminosity. Up it went, a great wall of fire about a mile in diameter, changing colors as it kept shooting upward, from deep purple to orange, expanding, growing bigger, rising as it was expanding, an elemental force freed from its bonds after being chained for billions of years.*

WILLIAM L. LAURENCE, *THE NEW YORK TIMES*, 26 SEPTEMBER 1945, ACCOUNT OF THE TRINITY TEST ON 16 JULY 1945

In Cold Blood

A rough guide to the symptoms resulting from different core body temperatures.

36–37°C
No symptoms – normal body temperature

35–36°C
Goosebumps, hands numb, breathing and pulse quickened

34–35°C
Movement slow, mild confusion

32–34°C
Shivering typically stops. Thinking slow and speech slurred

IDENTIFYING THE SIGNS OF HYPOTHERMIA

Normal body temperature is 37°C. Hypothermia – a condition where body temperature becomes dangerously low – is defined as a person having a core temperature below 35°C and is extremely serious. The list of signs and symptoms below is meant only as a general guide as there's a lot of variation between individuals. For example, consciousness can be lost after a drop of only 4°C in the core temperature, and even above 25°C death can occur.

30–32°C
Unable to walk, pupils dilated

28–30°C
Semi-conscious, slow pulse and irregular heartbeat

25.5–28°C
Unconscious

20°C
Heart stops

17°C
No electrical activity in the brain

13.7°C
Lowest recorded temperature that a person has recovered from

More than the Germ of an Idea

THE HISTORY OF PASTEURISATION

Louis Pasteur (1822–1895), perhaps France's greatest ever scientist, was a driving force in the development of the germ theory of disease. His interest in micro-organisms began during his work on fermentation, the study of which would eventually lead to the process bearing his name: pasteurisation. It was first used on wine, only later being adopted by milk producers.

Pasteur effectively ended the prevailing theory of spontaneous generation – the idea that living organisms can generate from inanimate objects – before turning his attention to the mechanism behind contagious diseases, in particular those afflicting silkworms in southern France, which had caused a crisis in the silk industry. At the same time, others were investigating the cause of anthrax, a disease deadly to both livestock and humans.

By the time Pasteur finished investigating silkworms in 1870, the idea that microbes were the cause of anthrax was hanging in the balance, with many scientists willing to dismiss the idea. The principal problem concerned how, if tiny organisms were responsible, outbreaks could suddenly occur in healthy livestock that had no contact with infected animals. It's at this point that the other outstanding name in the history of germ theory, the small-town German doctor Robert Koch, enters the story.

After meticulously studying the life cycle of anthrax, in 1876 Koch had the brilliant insight that spores of the disease could exist in the general environment for a very long time and *then* activate once inside an animal. He extended the scope of his studies

Louis Pasteur

Physicists use the wave theory on Mondays, Wednesdays and Fridays, and the particle theory on Tuesdays, Thursdays and Saturdays.

WILLIAM HENRY BRAGG (1862–1942), ENGLISH SCIENTIST AND CO-WINNER OF THE 1915 NOBEL PRIZE IN PHYSICS

and soon concluded that specific microbes caused specific diseases. This was another step altogether and the rest of the world needed convincing. A big boost to the argument was to come from the development of vaccines – tiny amounts of a disease that, if introduced into a body, could stimulate the production of disease-fighting antibodies.

Pasteur had been studying anthrax since 1879, and in the first half of 1881 he confidently announced that he and his team had managed to produce a vaccine for the disease and willingly accepted a scientific trial to publicly prove its efficacy. Taking place at the farm of one of his challengers, fifty healthy sheep were rounded up and half were injected with the vaccine. The remaining twenty-five were left as they were. Two weeks later, all fifty were injected with the microbe supposedly behind the disease. Within three days, all twenty-five 'control' sheep were dead, leaving only those that had been vaccinated alive. Played out so publicly, the demonstration was an extraordinary triumph and catapulted Pasteur into the scientific pantheon.

KOCH'S POSTULATES

Despite this success, however, the moves towards definitively proving that specific microbes caused specific diseases were essentially happening ad hoc, an inadequate method for settling that germ theory was correct once and for all. A more general approach was necessary. Here, we come back to Robert Koch. After his work on anthrax, Koch decided that what was needed was a formalised list of requirements, which, if met, would prove that a particular disease was caused by a particular micro-organism. This list is now known as Koch's postulates. He stipulated that:

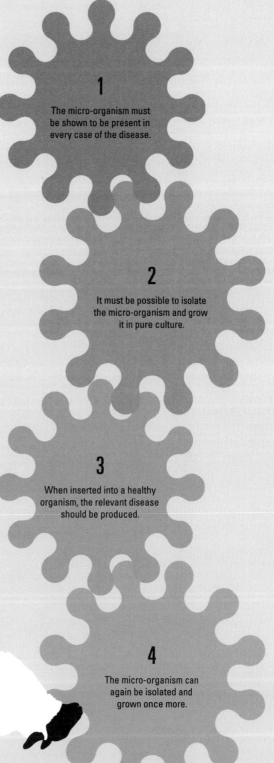

1
The micro-organism must be shown to be present in every case of the disease.

2
It must be possible to isolate the micro-organism and grow it in pure culture.

3
When inserted into a healthy organism, the relevant disease should be produced.

4
The micro-organism can again be isolated and grown once more.

From Greek Particles to Plum Puddings and Beyond

A SHORT HISTORY OF THE ATOM

The word 'atom' comes from the ancient Greek *atomos*, meaning 'uncuttable'. The idea of the existence of atoms was first proposed by the Greek philosophers Leucippus of Miletus (fl. 450–420 BC) and Democritus (c. 460–370 BC). They believed there were indivisible particles of matter, atoms, that moved around a void, and that it was the different patterns these were arranged in that produced change in the world, which we were then able to perceive.

As the English health journalist Andrew Gregory makes clear in his 2001 book *Eureka! The Birth of Science*, this distinction between the reality of the atomic world and how it then appears and is perceived by us is remarkable for its foresight. This original ancient view was refined and expanded by later thinkers and reached its apotheosis in the 1st-century BC poem *De rerum natura* (*On the Nature of Things*), by the Roman philosopher Lucretius, the rediscovery of which in the 15th century is the subject of the 2012 Pulitzer prize-winning book *The Swerve: How the Renaissance Began*, by American literary historian Stephen Greenblatt.

However, it wasn't until the very beginning of the 19th century, when English scientist John Dalton discovered atomic weight, that the concept of the atom could truly be said to have had practical consequences. Only from this point was there an understanding of chemical distinctness – the idea that each element is composed of its own atoms that in turn give the element its characteristic weight.

In 1897, English physicist J. J. Thomson announced the discovery of the electron. Less than ten years later, he discovered a correlation between the number of electrons in an atom and its atomic weight (in fact, the correlation was with an element's atomic number but hadn't yet been defined – it depends on the number of protons in an atom, and these had yet to be discovered). This led Thomson to propose the 'plum pudding' model of the atom, where the electrons are dotted throughout an atom, the rest of which is a kind of positively charged cloud, resulting in an overall neutral charge.

Believe it or not, the actual existence of the atom, as opposed to the attitude that it was merely a useful concept, wasn't accepted until 1905. It was then that Albert Einstein showed, using the relatively new mathematics of statistical mechanics, that atoms were responsible for Brownian motion – the random movement of particles suspended in a liquid. The phenomenon was named after Scottish botanist Robert Brown, who had noted the phenomenon more than seventy years earlier when observing pollen particles in water.

A 15-INCH SHELL

The plum pudding model of the atom didn't last long, though. In 1911, Thomson's former student Ernest Rutherford – a New Zealand-born British physicist – announced a new model after interpreting the startling results of an experiment performed two years earlier. In this experiment, a very small number of alpha particles had bounced back when fired at gold foil, something that shouldn't have happened if the plum pudding model was correct – if an atom was a kind of cloud, there should have been nothing for a particle to bounce off. As Rutherford memorably remarked, 'It was almost as incredible as if you had fired a 15-inch shell at a piece of tissue paper and it came back and hit you'. In his new model, the atom now contained a tiny positively charged nucleus, which was responsible for nearly all the atom's mass – it was when an alpha particle hit this that it rebounded. Electrons 'orbited' the nucleus like planets around the Sun. This is the view of the atom that we still learn at school.

One problem with Rutherford's model, however, was the question of electrons staying in orbit. According to classical mechanics, an electron should *gradually* lose energy, resulting in it spiralling into the nucleus. The answer of the Danish physicist Niels Bohr in 1913 was to quantise the atom. He argued that there existed only very specific energy levels an electron could occupy. A way of imagining this is to think of a tennis ball, which represents an electron, and a series of steps it can sit on, which represent the energy levels. The higher up the steps, the higher the energy and vice versa. In the atom, for an electron to move between these levels, energy, in the form of electromagnetic radiation, needs to be emitted or absorbed. There could be no gradual loss of energy and the spiral problem was solved. It was for this work that Bohr was awarded the Nobel Prize in Physics nine years later.

For centuries, a debate had raged as to whether light was a particle or a wave. By the beginning of the 20th century, it was generally accepted to be both. In 1924, not long after Bohr's Nobel Prize, the French physicist Louis de Broglie asserted in his PhD thesis that the electron was also guilty of this wave–particle duality. Austrian physicist Erwin Schrödinger subsequently realised that every wave had a corresponding equation describing it and began to work on one to describe de Broglie's discovery. This was revealed to the world in 1926 in the form of the Schrödinger equation, in which Ψ, the capital of the Greek letter *psi*, represents the wave function – something he felt was intimately connected with the cloud-like distribution of an electron's electric charge. This was because Schrödinger's equation concerned the probability distribution of a particle – how likely a particular position was. The fact it was now not possible to treat an electron as something for which the movement could be accurately traced (because the concept of probability had entered the mix) led the German physicist Werner Heisenberg in 1927 to formulate his 'uncertainty principle', in which it's impossible to know precisely both the position and the momentum (involving its velocity, mass and direction) of an electron. This put paid to the idea that if you knew the position and momentum of every particle in the universe, then their past and future could be known. Determinism was dead.

Q:

Why are quantum physicists so poor at sex?

A:

Because when they find the position, they can't find the momentum, and when they have the momentum, they can't find the position.

BOHR ATOMIC MODEL OF A CARBON ATOM

In a carbon atom, the nucleus comprises six protons and six neutrons. Surrounding the nucleus are six electrons – two on the first energy level and four on the second.

EXCITED ELECTRONS

When an electron changes energy level, energy is transferred via electromagnetic radiation.

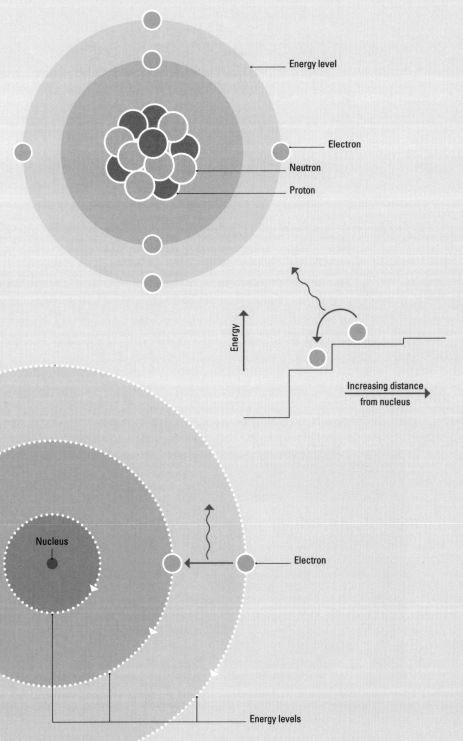

Energy level

Electron

Neutron

Proton

Energy

Increasing distance from nucleus

Nucleus

Electron

Energy levels

Letting the
Cat Out of the Bag

SCHRÖDINGER'S GROUNDBREAKING THOUGHT EXPERIMENT

In 1935, Austrian physicist Erwin Schrödinger published
what has become arguably the world's most famous thought
experiment. It involves a cat that is neither alive nor dead.
If that sounds almost nonsensical, in a way it's meant to – the
experiment is designed to draw attention to one of the strangest
elements of quantum theory, and, as the Danish physicist Niels
Bohr once remarked, 'Those who are not shocked when they first
come across quantum theory cannot possibly have understood it'.

At the 1927 Solvay Conference on Physics, held in Brussels, an attempt was made
to clarify the meaning of some recent studies in quantum theory. The view settled
on became known as 'the Copenhagen interpretation'. At its heart lies the role of
observation. In essence, Bohr and his followers'
version of quantum theory inferred that subatomic
events existed as probabilities that could be
resolved *only* when the event was observed in some
way. So, for example, a given particle might be in
any one of a number of positions within a locale,
there being a particular probability associated with
each position. It's the actual act of looking to check
where the particle is that then determines where
it's located. Let's just let the enormity of that sink
in for a moment – the idea is that reality
settles into a particular configuration
only when being looked at.

For Schrödinger, the
implied effect this sense of
being in limbo at the

Erwin Schrödinger

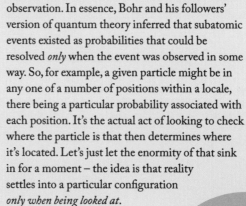

'Why,' said the
Dodo, 'the best way to
explain it is to do it.'
**LEWIS CARROLL (1832–1898),
ENGLISH WRITER AND
MATHEMATICIAN,
*ALICE'S ADVENTURES IN
WONDERLAND* (1865)**

subatomic level had on our everyday, macroscopic world was simply unacceptable. In response to correspondence with Einstein, who egged him on, Schrödinger's thought experiment was intended to highlight this absurdity. It goes like this:

> A cat is penned up in a steel chamber, along with the following diabolical device (which must be secured against direct interference by the cat): in a Geiger counter there is a tiny bit of radioactive substance, so small that perhaps in the course of one hour, one of the atoms decays, but also, with equal probability, perhaps none; if it happens, the counter tube discharges and through a relay releases a hammer, which shatters a small flask of hydrocyanic acid. If one has left this entire system to itself for an hour, one would say that the cat still lives if meanwhile no atom has decayed. The first atomic decay would have poisoned it. The wave function of the entire system would express this by having in it the living and the dead cat (pardon the expression) mixed or smeared out in equal parts.

Bohr was unrepentant, refusing to accept the thought experiment's conclusion and insisting the cat would be either dead or alive before any observation was made. It's perhaps an irony, given its purpose was to ridicule Bohr's views, that Schrödinger's cat has had the impact it has in the world beyond science. Literature, in particular, has mined this paradox, including Ursula le Guin's irreverent short story 'Schrödinger's Cat' and Frederik Pohl's science fiction novel *The Coming of the Quantum Cats*. Perhaps of lower aesthetic merit, but no less entertaining, it has also inspired some tongue-in-cheek poetry, such as the following by Marilyn T. Kocher:

'The Tale of Schrödinger's Cat'

Schrödinger called his cat and said,
'You can be both alive and dead,
For a linear combination of states
Postulates two simultaneous fates.'
Poor shocked pussy could not say,
'I shall inform the SPCA.
Your pet theory seems to me
An ultraviolent catastrophy.'

What then did this kitty do?
She looked at him and said 'μ'.

'Scientists Anonymous'

INFLUENTIAL FEMALE SCIENTISTS

Open a typical book on the history of science and you'll find references to few female scientists. There are many reasons for this, but two stand out: a historical lack of opportunity for women in science and deeply ingrained prejudice. Yet, despite these considerable challenges, women made numerous contributions to the growth of scientific knowledge over the centuries. Here are some of my favourites, along with a snapshot of their achievements:

1706–1749

ÉMILIE DU CHÂTELET

Translated Sir Isaac Newton's *Principia* into French and included a commentary that clarified and, at times, extended Newton's magnum opus and featured her derivation of the principle of energy conservation.

1769–1858

JANE MARCET

The author of *Conversations on Chemistry, Intended More Especially for the Female Sex* (1805), which went through seventeen editions and introduced the bookbinder's apprentice Michael Faraday to electrochemistry.

1711–1778

LAURA BASSI

The first woman to become a professor in Europe, Bassi was an expert on Isaac Newton's work and played a key role in communicating his ideas.

1799–1847

MARY ANNING

Discovered several fossil firsts, including a complete Plesiosaurus and a British Pterodactylus.

1758–1836

MARIE-ANNE PAULZE LAVOISIER

Made clear and detailed illustrations relevant to her husband Antoine Lavoisier's chemical work that helped spread the understanding of his work.

1818–1889

MARIA MITCHELL

Became the first woman elected to the American Academy of Arts and Sciences, a year after she discovered the comet C/1847 T1. She later became the first female professional astronomer in the USA.

1850–1891

SOFIA KOVALEVSKAIA

Solved the structure of Saturn's rings and extended our understanding of calculus. She was the first woman to be awarded a doctorate.

1867–1934

MARIE CURIE

Probably the best-known female scientist, Curie was the first woman to win a Noble Prize and the first person to win it twice – one for her work on radioactivity and the other for the discovery of radium and polonium – and her subsequent investigations into the former.

1910–1994

DOROTHY HODGKIN

Only the third female winner of the Nobel Prize for Chemistry, Hodgkin determined the structure of vitamin B_{12}, a molecule containing more than 100 atoms, and, later on, the protein insulin.

1854–1923

HERTHA AYRTON

Born Phoebe Sarah Marks, Ayrton was a leading expert on the electric arc and the only female member of the Institution of Electrical Engineers until 1958.

1878–1968

LISE MEITNER

Co-discoverer of fission that ushered in the nuclear age and furthered the development of radioactivity.

1917–1966

CAROLYN PARKER

Believed to be the first African American to be awarded a postgraduate degree in physics, Parker worked on a part of the Manhattan Project that focused on the radioactive element polonium.

1863–1914

IDA FREUND

Director of Studies of Chemistry and Physics at Newnham College, Cambridge, Freund was a major force behind a quest to improve science teaching in girls' schools in the UK.

1892–1916

ALICE BALL

The first African American and woman to be awarded a Master's degree from the University of Hawaii, Ball developed a drug to treat leprosy. She died aged just 24.

1920–1958

ROSALIND FRANKLIN

X-ray crystallographer who took the all-important Photograph 51 of DNA, which proved to be a crucial step in understanding its double-helix structure.

Truth to Their Fictions

SCIENTISTS IN WRITING

Given the large number of people who work in science, it's perhaps surprising that more scientists don't appear in fiction. In part inspired by the website www.lablit.com, which is dedicated to promoting 'the culture of science in fiction and fact', here's a list of titles in which they do.

For better or worse, the list excludes any title that might be considered science fiction, such as Aldous Huxley's *Brave New World*, in an attempt to ground the list as much as possible in the 'real world'. I appreciate I haven't necessarily wholly succeeded in achieving this.

Brazzaville Beach, William Boyd

In this book, an ecologist travels to Africa to study primates and escape her marriage to a brilliant but flawed mathematician. It won the McVitie's and James Tate Black Memorial prizes in 1990.

The Gold Bug Variations, Richard Powers

Named by *Time* magazine as the best novel of 1991, this novel of ideas, with two love stories at its heart, draws on a variety of disciplines, including genetics, computer programming, Flemish art and music composition. Powers is also the author of *The Overstory* and *Bewilderment* (shortlisted for the Booker Prize in 2021), both of which are relevant to this list.

Einstein's Dreams, Alan Lightman

Sometimes compared to Italo Calvino's *Invisible Cities* and Jorge Luis Borges's *Labyrinths*, this series of connected fables reveals the dreams Albert Einstein might have had when finalising his special theory of relativity in 1905.

Measuring the World, Daniel Kehlmann

Originally published in German, and an international bestseller, this is a good-humoured fictional account of a meeting between two German giants of 18th- and 19th-century intellectual thought, the naturalist Alexander von Humboldt and the mathematician Carl Friedrich Gauss.

This Thing of Darkness, Harry Thompson

Longlisted for the Booker Prize in 2005, this retelling of the famous voyage of HMS *Beagle*, and the following thirty years, focuses on the relationship between the ship's captain, Robert Fitzroy, and its scientific passenger, Charles Darwin.

Ship Fever, Andrea Barrett

A National Book Award winner in 1996, this collection of short stories set in the 19th century takes its inspiration from the practice of science, humanising it in the process.

Remarkable Creatures, Tracy Chevalier

A novel that brings to life Mary Anning, an early 19th-century collector and seller of fossils in Lyme Regis, in Dorset, England, whose discoveries helped to shape the new science of palaeontology but who, due to her sex, received little credit for her finds.

Solar, Ian McEwan

A satirical novel on climate change written by an author who draws on science for many of his novels. It won the Bollinger Everyman Wodehouse Prize for comic fiction in 2010.

Thinks, David Lodge

A clever campus novel that examines the puzzle that is consciousness, through the eyes of a male cognitive scientist and a female novelist who embark upon an illicit affair.

Two on a Tower, Thomas Hardy

Covering the relationship between the wife of a country squire and a poor astronomer ten years her junior, Hardy wanted to 'set the emotional history of two infinitesimal lives against the stupendous background of the stellar universe'.

The Housekeeper and the Professor, Yoko Ogawa

This short novel tells the moving story of a young mother employed to look after a mathematician who years earlier suffered a head injury. As a result, he can remember everything before the accident but can't hold on to anything since for more than eighty minutes. Every day, the housekeeper must start from scratch but soon learns to speak to him in the language of maths.

The Calcutta Chromosome, Amitav Ghosh

An imaginative tale that switches between the late 19th century and British doctor Ronald Ross's work on the transmission of malaria and the modern day, where a Ross enthusiast is desperate to get to the truth – which turns out to be stranger than fiction – behind how the Nobel Prize-winning discovery was made.

The New Men, C. P. Snow

The sixth book in Snow's *Strangers and Brothers* series, this 1954 novel centres on a group of scientists striving to produce controlled nuclear fission during the Second World War and the government's management of them.

Arrowsmith, Sinclair Lewis

Winner of the Pulitzer Prize in 1926, which the author declined, this story relates the journey through life of Martin Arrowsmith, from his humble beginnings in the Midwest to scientific success despite the death of his wife from a plague for which he eventually finds a cure.

Cantor's Dilemma, Carl Djerassi

By a renowned professor of chemistry, this novel delves into the murky high-stakes politics of scientific research and explores the question of just how far someone would go to secure a Nobel Prize.

Kepler, John Banville

The second book in his *Revolutions* trilogy, here Banville recreates the life of the astronomer Johannes Kepler, casting light on the period he lived in and his quest to understand the universe.

The Mind–Body Problem, Rebecca Goldstein

In this book, a beautiful philosophy graduate rebelling against her strict
orthodox Jewish upbringing finds herself adrift in life, eventually marrying a
brilliant Princeton mathematician through a desire for reflected status.
However, their respective world views clash, mainly due to their positions on
the centuries-old mind–body problem (the difficulty of understanding the
connection between our physical bodies and our mental worlds).

The Indian Clerk, David Leavitt

Based on the relationship between the respected British mathematician
G. H. Hardy and the autodidact mathematical genius Srinivasa Ramanujan
who, while a clerk in Madras in 1913, was invited to Cambridge by Hardy,
who had read samples of his brilliant theorems.

Experimental Heart, Jennifer Rohn

By the founder of www.lablit.com, this contemporary romantic thriller
focuses on two laboratory-based scientists and the external pressure upon
them to achieve results.

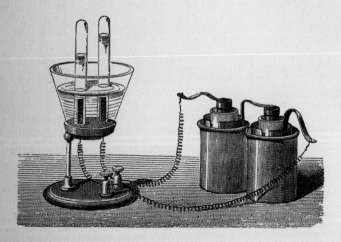

The Life of
π

BEYOND THE DECIMAL POINT OF 3.14 ...

Leonhard Euler introduced the Greek letter π to denote the ratio of a circle's circumference to its diameter during the 18th century. The concept already had an illustrious history by that point, and many ancient civilisations had determined useful approximations with which to measure π…

ANCIENT PI

The Babylonians give a ratio of $3\frac{1}{8}$ and the Egyptians a ratio of $^{256}/_{81}$.

HOLY PIBLE

The Bible suggests a ratio of 3 in Kings 7:23.

GREEK PI

The ancient Greeks came up with a variety of ratios, the best being the 2nd-century mathematician Hipparchus's value of $^{377}/_{120}$, which is correct to four decimal places.

PI IN ASIA

During the first millennium, astronomers in China and India suggested values correct to between three and six decimal places, including Arybatha in AD 499 and Tsu Chúng-chih (AD 430–501).

ADVANCED PI

It wasn't until 1424, when Jamshid Masud al-Kashi published *al-Risala al-Muhitiyya*, that a significant advance in the calculation of π was made – correct to sixteen decimal places.

EUROPEAN PI

Europe began to make a more serious effort towards the end of the 16th century, when the race to calculate π properly started.

Pi, represented by the Greek letter π, is a mathematical constant calculated by dividing the circumference (C) of a circle by its diameter (d).

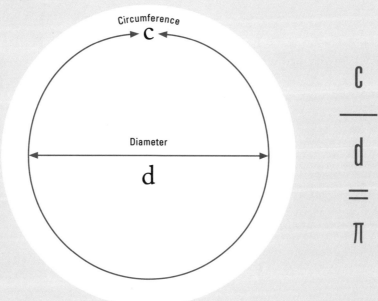

Circumference

c

Diameter

d

$$\frac{c}{d} = \pi$$

INFINITE PI

To anyone who has the time and the inclination, there is now no real maximum to the number of decimal places possible, thanks to the modern, pi-throwing computer.

In case you're hungry for it, here's a large piece of pi:

3.1415926535897932384626433832795028841
9716939937510582097494459230781640628
20899862803482534211706798214808651328
23066470938446095505822317253594081284
8111745028410270193852110555964462294
9549303819644288109756659334461284756
823378678316527120190914564856692346034
48610454326648213393607260249141273724
58700660631558817488152092096282925409

A Slice of Indiana π

AN ATTEMPT TO MONETISE PI

For more than 2,000 years, people were determined to crack the problem of squaring the circle, that is, constructing a square the area of which is *exactly* the same as that of a given circle, using a straight edge and compass. Hopes that it might be possible were quashed when, in 1882, German mathematician Ferdinand von Lindemann showed that π was, in fact, a transcendental number, meaning it wasn't the root of any algebraic equation. But this didn't put off the American physician Edwin J. Goodwin who, in 1897, wrote a bill for the state of Indiana with the intention of establishing a 'new mathematical truth' – his procedure for squaring the circle.

THE LUNE OF HIPPOCRATES
Hippocrates of Chios discovered that these shaded areas are of the same size. This offered false hope of squaring the circle.

On 18 January 1897, House Bill 246 was submitted to the Indiana General Assembly. Amazingly, Goodwin wanted to copyright his solution, and therefore make money from it, declaring in the opening of the bill that it was to be 'offered as a contribution to education to be used only by the State of Indiana free of cost by paying any royalties whatever on the same, provided it is accepted and adopted by the official action of the Legislature of 1897'.

There was just one major flaw in Goodwin's new truth. Tucked away in the second section of the Bill was the 'fourth important fact' necessary to the method, that 'the ratio of the diameter and circumference is as five-fourths to four' or, rather, 3.2. So, Goodwin was essentially trying to establish in State law that π = 3.2, rather than 3.1459 … and then to make money from this 'fact'!

Astonishingly, the bill was passed on 5 February by sixty-seven votes to none. It was then transferred to the Indiana Senate where its first reading by the Committee on Temperance was reported back favourably and put formally before the Senate on 12 February. Thankfully, common sense was given a serendipitous opportunity – Professor Clarence Abiathar Waldo, a mathematician at Purdue University, had been visiting the Statehouse and overheard the General Assembly debate. He 'coached' the senators about the Bill, which was given short shrift and postponed indefinitely. This rather suggests it's still knocking about somewhere.

> Where do you bury dead mathematicians?
> Asymmetry

A Monk Who Liked His Peas

MENDEL AND THE FOUNDATIONS OF GENETICS

Born in Silesia, part of the Austrian Empire, Gregor Johann Mendel (1822–1884) was a monk whose experimental work laid the foundations of genetics. In 1866, he published the results of thousands of experiments in breeding peas, which he'd carried out in the gardens of his monastery. Mendel discovered there was a pattern discernible in the inheritance of characteristics such as flower colour in pea plants. This led him to formulate his law of segregation.

The law states that two units control the inheritance pattern of a particular characteristic in an organism. We now know these to be alleles, or different forms of a gene. An organism will carry two forms of the same gene, one from each of its parents. When germ (sex) cells are formed, they receive only one of this pair. So, because they're made by the fusion of a germ cell from each parent, the first offspring generation (the first filial generation) contains an allele from each parent. Therefore, if one of the parent plants has pure white flowers and the other pure purple flowers, the first filial generation will carry one allele for purple flowers and one for white flowers.

With the breeding of successive generations, it became clear to Mendel that certain traits could be either dominant or recessive and that the dominant would always mask the recessive. This explained particular patterns of inheritance and why traits could skip generations. For example, if we look at the inheritance of the relevant colour allele when pure purple and pure white pea plants are crossed, we notice the colour white disappears in the second generation, because purple is the dominant trait. But white comes back in the third generation, because it's possible for a plant to inherit two white alleles at this stage (below).

Charles Darwin may have provided the theory of evolution with the sound footing it needed, but it was Gregor Mendel's work that enabled the study of the mechanisms that lay behind Darwin's theories.

1st generation

A   A   a   a

2nd generation

A   a   A   a   A   a   A   a

3rd generation

A   A   a   A   A   a   a   a

Mendel's Law in Action

In this example, A represents the allele for purple colouring and is dominant. Only when a plant inherits two recessive alleles does white colouring appear.

Meet Tom Telescope and Friends

NEWBERY'S SIX LECTURES

The John Newbery Medal is the most prestigious children's book prize in the United States today. Begun in 1922, and the first of its type, it's awarded by the American Library Association 'to the author of the most distinguished contribution to American literature for children' and named after the 18th-century English publisher and bookseller John Newbery, widely recognised as the first person in his field to specialise in children's books.

In 1761, Newbery published *The Newtonian System of Philosophy Adapted to the Capacities of Young Gentlemen and Ladies*, a popular science book probably aimed at children aged between ten and twelve. In the book, 'six Lectures [are] read to the Lilliputian Society, by Tom Telescope', which have been 'collected and methodized for the Benefit of the Youth of these Kingdoms'.

The lectures cover the full range of 18th-century natural philosophy, from matter and gravity to the five senses of man, and continually refer to objects with which young people would have been acquainted, such as balls and tops. What follows is the beginning of the first lecture, in which Newton's first law of motion is introduced very clearly and concisely:

> *When a body is in motion, as much force is required to make it rest, as was required, while it was at rest, to put it in motion. Thus, suppose a boy strikes a trap-ball with one hand, and another stands close by to catch it with one of his hands, it will require as much strength or force to stop that ball, or put it in a state of rest, as the other gave to put it in motion; allowing for the distance the two boys stand apart.*
>
> *No body or part of matter can give itself either motion or rest: and therefore a body at rest will remain so for ever, unless it be put in motion by some external cause; and a body in motion will move for ever, unless some external cause stops it.*
>
> *This seemed so absurd to Master Wilson, that he burst into a loud laugh. What, says he, shall any body tell me that my hoop or my top will run for ever, when I know by daily experience that they drop of themselves, without being touched by any body? At which our little Philosopher was angry, and having commanded silence, Don't expose your ignorance, Tom Wilson, for the sake of a laugh, says he: if you intend to go through my Course of Philosophy, and to make yourself acquainted with the nature of things, you must prepare to hear what is more extraordinary than this. When you say that nothing touched the top or the hoop, you forget their friction or rubbing against the ground they run upon, and the resistance they meet with from the air in their course, which is very considerable though it has escaped your notice. Somewhat too might be said on the gravity and attraction between the top or the hoop, and the earth; but that you are not yet able to comprehend, and therefore we shall proceed in our Lecture.*

Remind me not to get on the wrong side of Tom Telescope.

The book was a huge success, with numerous editions, and as well as being published in Ireland and America, was translated into Dutch, Swedish and Italian. Many historians now feel it highly likely that John Newbery himself wrote the book. It was almost certainly publications like this that prompted English essayist Charles Lamb – apparently not much of a fan of science – to write to the poet Samuel Taylor Coleridge in 1802 that 'Science has succeeded to poetry, no less in the little walks of children than with men. Is there no possibility of averting this sore evil?' If he were still alive today, Lamb might feel that both science and poetry have been 'averted', certainly as far as children are concerned.

The Sins of a Fledgling Scientist

INVESTIGATING ISAAC NEWTON'S RELIGIOSITY

Isaac Newton, widely regarded as one of the greatest scientists to have ever lived, was a deeply religious man. An anti-Trinitarian, and so an opponent of the Christian belief in the triple Godhead of Father, Son and Holy Spirit – a heretical position during the 17th century – Newton spent a huge amount of time and effort attempting to uncover secret knowledge in the sacred texts of ancient cultures. This included devising rules for interpreting the symbols of biblical prophecy – he was particularly interested in the books of Daniel and Revelations. Based on time periods in the former, Newton calculated the end of the world – the apocalypse – would not happen before 2060. We've a few years left, at least.

Isaac Newton

More mundanely, although certainly not for Newton himself, as a nineteen-year-old in 1662 he recorded in a notebook forty-eight sins he could recall committing up to Whitsunday. It makes fascinating reading and helps to turn a scientific giant into someone very human. It also provides an insight into Newton's religious fervour – something that never left him. The following is a selection from the list:

- Using the word (God) openly
- Making a feather while on Thy day
- Denying that I made it
- Squirting water on Thy day
- Making pies on Sunday night
- Threatning my father and mother Smith to burne them and the house over them
- Wishing death and hoping it to some
- Having uncleane thoughts words and actions and dreamese
- Stealing cherry cobs from Eduard Storer
- Denying that I did so
- Setting my heart on money, learning, pleasure more than Thee
- Punching my sister
- Robbing my mother's box of plums and sugar
- Calling Dorothy Rose a jade
- Peevishness with my mother
- Idle discourse on Thy day and at other times
- Not loving Thee for Thy goodness to us
- Not desiring Thy ordinances
- Fearing man above Thee
- Using unlawful means to bring us out of distresses
- Not craving a blessing from God on our honest endeavors
- Missing chapel
- Beating Arthur Storer
- Peevishness at Master Clarks for a piece of bread and butter
- Striving to cheat with a brass halfe crowne
- Twisting a cord on Sunday morning

From Newton's monument, Westminster Abbey:

H. S. E. ISAACUS NEWTON Eques Auratus, / Qui, animi vi prope divinâ, / Planetarum Motus, Figuras, / Cometarum semitas, Oceanique Aestus. Suâ Mathesi facem praeferente / Primus demonstravit: / Radiorum Lucis dissimilitudines, / Colorumque inde nascentium proprietates, / Quas nemo antea vel suspicatus erat, pervestigavit. / Naturae, Antiquitatis, S. Scripturae, / Sedulus, sagax, fidus Interpres / Dei O. M. Majestatem Philosophiâ asseruit, / Evangelij Simplicitatem Moribus expressit. / Sibi gratulentur Mortales, / Tale tantumque exstitisse / HUMANI GENERIS DECUS. / NAT. XXV DEC. A.D. MDCXLII. OBIIT. XX. MAR. MDCCXXVI

This can be translated as follows:

Here is buried Isaac Newton, Knight, who by a strength of mind almost divine, and mathematical principles peculiarly his own, explored the course and figures of the planets, the paths of comets, the tides of the sea, the dissimilarities in rays of light and, what no other scholar has previously imagined, the properties of the colours thus produced. Diligent, sagacious and faithful, in his expositions of nature, antiquity and the holy Scriptures, he vindicated by his philosophy the majesty of God mighty and good and expressed the simplicity of the Gospel in his manners. Mortals rejoice that there has existed such and so great an ornament of the human race! He was born on 25 December 1642, and died on 20 March 1726.

The Monkey Trial

EVOLUTION IN THE DOCK

On 13 March 1925, the State of Tennessee passed an incredible law – the Butler Act – that stated evolution could not be taught in any university or public school for which they provided funding. Anyone caught doing so faced a fine of up to $500, approximately half a teacher's annual salary. If anything, this felt like a challenge. The American Civil Liberties Union advertised for a teacher willing to make it known they had contravened the law and taught the theory of evolution. John T. Scopes was the man who stepped forward.

Thus began the Scopes Trial (also known as the Monkey Trial), one of the most famous court cases centring on science in the 20th century. It pitted the former US Secretary of State, William Jennings Bryan, acting for the prosecution, against the acclaimed lawyer Clarence Darrow. It was a trial that would be transmitted across America by radio and, ironically, provided an excellent opportunity for the theory of evolution to be explained to millions of people.

On the eighth day of the trial, after the majority of the defence's arguments had been declared inadmissible by a clearly biased judge, the defence changed their plea to guilty and Scopes was fined $100. This gave the defence team the opportunity to take the case to the Tennessee Supreme Court where, two years later, the validity of the Butler Act was upheld but Scopes's sentence was reversed on a technicality – the minimum fine, which had been enforced, was double what could be imposed without a jury's assessment and the original 'trial judge [had] exceeded his jurisdiction in levying this fine'. The Act, which would not be repealed for another forty years, is reproduced opposite:

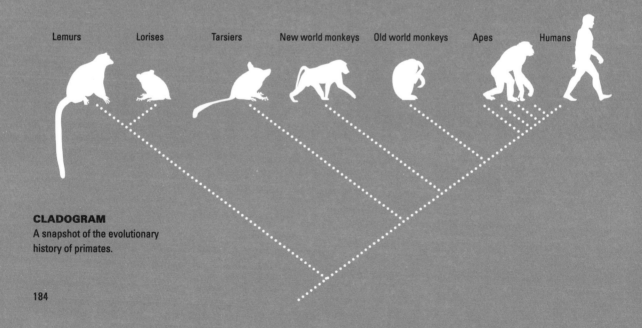

Lemurs　　Lorises　　Tarsiers　　New world monkeys　　Old world monkeys　　Apes　　Humans

CLADOGRAM
A snapshot of the evolutionary history of primates.

The Butler Act
House Bill No. 185

AN ACT prohibiting the teaching of the Evolution Theory in all the Universities, Normals and all other public schools of Tennessee, which are supported in whole or in part by the public school funds of the State, and to provide penalties for the violations thereof.

Section 1. *Be it enacted by the General Assembly of the State of Tennessee*, That it shall be unlawful for any teacher in any of the Universities, Normals and all other public schools of the State which are supported in whole or in part by the public school funds of the State, to teach any theory that denies the story of the Divine Creation of man as taught in the Bible, and to teach instead that man has descended from a lower order of animals.

Section 2. *Be it further enacted*, That any teacher found guilty of the violation of this Act, Shall be guilty of a misdemeanor and upon conviction, shall be fined not less than One Hundred ($100.00) Dollars nor more than Five Hundred ($500.00) Dollars for each offense.

Section 3. *Be it further enacted*, That this Act take effect from and after its passage, the public welfare requiring it. Passed March 13, 1925 (Repealed 1 September 1967).

What if Earth Can Clothe and Feed Amplest Millions at Their Need

A CHEERLEADER FOR SCIENCE

The radical English poet Percy Bysshe Shelley fought against the system throughout his short, twenty-nine-year life, which ended in 1822. At Eton College, he wouldn't 'fag' – a system where a younger boy effectively serves an older one – and was ostracised by his fellow pupils. At Oxford, he lasted only a term after writing and then refusing to withdraw 'The Necessity of Atheism', a pamphlet he co-wrote with his college friend, the future barrister and writer Thomas Jefferson Hogg, which had been sent to every head of college.

A vociferous and varied reader, Shelley was a thinker of extraordinary ideas and it's hard to find a poet more enamoured with science and the promise it held for improving the lot of the general populace. Shelley's notes on Humphry Davy's *Elements of Agricultural Chemistry* (1813) alone ran to more than twenty pages, and he possessed all manner of scientific equipment in his college rooms, including an electrical machine and an air pump.

Hogg wrote about his time with Shelley at Oxford in *The New Monthly Magazine and Literary Journal* just over twenty years later, a decade after Shelley's premature death. In it is a passage that relates Shelley's belief that science could benefit society, improving the welfare of everyone, particularly the impoverished. Even considering the second-hand nature of the account, and while still trying to be careful about reading predictions made 200 years ago with today's eyes, Shelley's thoughts are nevertheless remarkable. Allowing for the odd tweak here and there, you could argue that nearly every single one has come true. It's the optimism science can provide, as articulated by Shelley, that makes it as important and as wonderful as it is. It just needs to be handled in the way Shelley imagines.

'Is not the time of by far the larger proportion of the human species', he inquired, with his fervid manner and in his piercing tones, 'wholly consumed in severe labour? And is not this devotion of … the whole of our race (for those who … are indulged with an exemption from the hard lot are so few … that they scarcely deserve to be taken into the account) absolutely necessary to procure subsistence; so that men have no leisure for recreation or the high improvement of the mind? Yet this incessant toil is still inadequate to procure an abundant supply of the common necessaries of life: some are doomed actually to want them, and many are compelled to be content with an insufficient provision.

'We know little of the peculiar nature of those substances which are proper for the nourishment of animals; we are ignorant of the qualities that make them fit for this end. Analysis has advanced so rapidly of late that we may confidently anticipate that we shall soon discover wherein their aptitude really consists; having ascertained the cause, we shall next be able to command it, and to produce at our pleasure the desired effects.

'It is easy, even in our present state of ignorance, to reduce our ordinary food to carbon, or to lime; a moderate advancement in chemical science will speedily enable us, we may hope, to create, with equal facility, food from substances that appear at present to be as ill adapted to sustain us.

'What is the cause of the remarkable fertility of some lands, and of the hopeless sterility of others? A spadeful of the most productive soil does not to the eye differ much from the same quantity taken from the most barren. The real difference is probably very slight; by chemical agency the philosopher may work a total change, and may transmute an unfruitful region into a land of exuberant plenty.

'Water, like the atmospheric air, is compounded of certain gases: in the progress of scientific discovery a simple and sure method of manufacturing the useful fluid, in every situation and in any quantity, may be detected; the arid deserts of Africa may then be refreshed by a copious supply, and may be transformed at once into rich meadows, and vast fields of maize and rice.

'The generation of heat is a mystery, but enough of the theory of caloric has already been developed to induce us to acquiesce in the notion that it will hereafter, and perhaps at no very distant period, be possible to produce heat at will, and to warm the most ungenial climates as readily as we now raise the temperature of our apartments to whatever degree we may deem agreeable or salutary. If, however, it be too much to anticipate that we shall ever become sufficiently skilful to command such a prodigious supply of heat, we may expect, without the fear of disappointment, soon to understand its nature and the causes of combustion, so far at least as to provide ourselves cheaply with a fund of heat that will supersede our costly and inconvenient

fuel, and will suffice to warm our habitations, for culinary purposes and for the various demands of the mechanical arts.

'We could not determine, without actual experiment, whether an unknown substance were combustible; when we shall have thoroughly investigated the properties of fire, it may be that we shall be qualified to communicate to clay, to stones, and to water itself, a chemical recomposition that will render them as inflammable as wood, coals, and oil, for the difference of structure is minute and invisible, and the power of feeding flame may perhaps be easily added to any substance, or taken away from it. What a comfort would it be to the poor at all times, and especially at this season, if we were capable of solving this problem alone, if we could furnish them with a competent supply of heat!

'These speculations may appear wild, and it may seem improbable that they will ever be realised, to persons who have not extended their views of what is practicable by closely watching science in its course onward; but there are many mysterious powers, many irresistible agents, with the existence and with some of the phenomena of which all are acquainted. What a mighty instrument would electricity be in the hands of him who knew how to wield it, in what manner to direct its omnipotent energies; and we may command an indefinite quantity of the fluid: by means of electrical kites we may draw down the lightning from heaven! What a terrible organ would the supernal shock prove, if we were able to guide it; how many of the secrets of nature would such a stupendous force unlock! The galvanic battery is a new engine; it has been used hitherto to an insignificant extent, yet has it wrought wonders already; what will not an extraordinary combination of troughs, of colossal magnitude, a well-arranged system of hundreds of metallic plates, effect? The balloon has not yet received the perfection of which it is surely capable; the art of navigating the air is in its first and most helpless infancy; the aerial mariner still swims on bladders, and has not mounted even the rude raft: if we weigh this invention, curious as it is, with some of the subjects I have mentioned, it will seem trifling, no doubt a mere toy, a feather, in comparison with the splendid anticipations of the philosophical chemist; yet it ought not altogether to be condemned. It promises prodigious facilities for locomotion, and will enable us to traverse vast tracts with ease and rapidity, and to explore unknown countries without difficulty. Why are we still so ignorant of the interior of Africa? Why do we not despatch intrepid aeronauts to cross it in every direction, and to survey the whole peninsula in a few weeks? The shadow of the first balloon, which a vertical sun would project precisely underneath it, as it glided silently over that hitherto unhappy country, would virtually emancipate every slave, and would annihilate slavery for ever.'

Five, Four, Three, Two, One …
We Have Lift-off!

THE SCIENCE OF SPACE SHUTTLE FLIGHT

From launch, every space shuttle flight took just nine minutes to attain orbit and a speed of more than 28,000km/h (17,400mph) – more than twenty-two times the speed of sound. Helping to achieve this were three very carefully thought-out chemical reactions.

The two rocket boosters attached to the shuttle's external fuel tank were chiefly responsible for powering the first two minutes of the craft's journey – that's about the first 45km (28 miles). At the point of lift-off, the total mass of the shuttle was 2,000 tonnes, so an incredibly large lifting force was needed to start it on its travels. The boosters contained a mixture of the fuel (powdered aluminium), a powerful oxidiser (ammonium perchlorate) and a small amount of catalyst (iron (III) oxide) to help the reaction along. These reacted together to produce steam and nitric oxide that, due to the temperature inside the boosters reaching more than 3,000°C from the heat of the reaction, expanded extremely rapidly to create huge propulsive force. Also produced were solid aluminium oxide and aluminium chloride, which could be seen as the dense white clouds expelled from the rockets.

$$3Al_{(s)} + 3NH_4ClO_{4(s)} \rightarrow Al_2O_{3(s)} + AlCl_{3(s)} + 6H_2O_{(g)} + 3NO_{(g)}$$

Once the boosters were jettisoned, the main engines on the shuttle itself became responsible for getting it into orbit. For the next six minutes, these used liquid hydrogen and liquid oxygen stored in the external fuel tank. The resulting water vapour was expelled from the engines at almost 10,000km/h (6,200mph).

$$2H_{2(l)} + O_{2(l)} \rightarrow 2H_2O_{(g)}$$

The final propulsion system was used to manoeuvre the shuttle once it was in orbit. It was vital that this stage involved a reaction that could be started and stopped exactly as required. Here, the fuel was monomethylhydrazine and the oxidiser dinitrogen tetraoxide. This again produced water vapour, but also carbon dioxide.

$$CH_3(NH)NH_2 + 5N_2O_{4(l)} \rightarrow 12H_2O_{(g)} + 4CO_{2(g)} + 9N_{2(g)}$$

The scientist must set in order. Science is built up with facts, as a house is with stones. But a collection of facts is no more a science than a heap of stones is a house.

HENRI POINCARÉ (1854–1912), FRENCH MATHEMATICIAN, PHYSICIST AND PHILOSOPHER, *SCIENCE AND HYPOTHESIS* (1902)

Blowing Hot and Cold

184K
Lowest temperature recorded on
Earth, in Antarctica

HEATING UP
A temperature scale in Kelvin,
from absolute zero to the searing
heat of the surface of the Sun.

195K
Sublimation temperature of dry ice (carbon dioxide)

261K
Highest recorded temperature at the South Pole

20K
Boiling point
of hydrogen

298K
Room temperature

77K
Boiling point of
nitrogen

373K
Boiling point
of water

OK	100K	200K	300K	400K	500K	600K	700K

Absolute
zero

330K
Hottest temperature
recorded on Earth,
in Libya

601K
Melting point of lead

226K
Average
temperature
on Mars

506K
Auto-ignition point
of paper according
to Ray Bradbury's
novel *Fahrenheit 451*

2.73K
Average temperature of the universe
(cosmic background radiation)

733K
Average temperature on Venus

1600K
Temperature of
space shuttle
re-entry

1811K
Melting point of iron

5780K
Surface of
the Sun

800K   900K   1000K   2000K   3000K   4000K   5000K   6000K

1337K
Melting point
of gold

*A first-rate
theory predicts;
a second-rate theory forbids;
and a third-rate theory explains
after the event.*

**ALEKSANDR ISAAKOVICH KITAIGORODSKII
(1914–1985), RUSSIAN PHYSICIST AND
CRYSTALLOGRAPHER, AT A LECTURE
AT IUC AMSTERDAM,
AUGUST 1975**

There Is More to Seeing than Meets the Eyeball

THE THEORETICAL
HORSE BEFORE THE FACTUAL CART?

Although extreme in his nature, Thomas Gradgrind – a school superintendent in Charles Dickens's novel *Hard Times* and the voice of the quotation on the opposite page – symbolises perfectly a particular view of science as something cold and clinical, involving absolute objectivity.

For many years, scientific theories were considered as something derived purely from facts – crudely put, it was believed that the more facts you had, or examined, the better your theory was likely to be. However, as someone confused aptly put it, 'the trouble with facts is that there are so many of them' – so, how are you to decide which of them are relevant?

Now, what I want is, Facts. Teach these boys and girls nothing but Facts. Facts alone are wanted in life. Plant nothing else, and root out everything else. You can only form the minds of reasoning animals upon Facts: nothing else will ever be of any service to them. This is the principle on which I bring up my own children, and this is the principle on which I bring up these children. Stick to Facts, sir!

THOMAS GRADGRIND IN *HARD TIMES* (1854), BY CHARLES DICKENS

There's an even bigger problem with a fact-leading-to-theory view of science – no matter how hard you try, it's pretty much impossible for observation, typically the source of a scientist's facts, not to be in some way theory-laden. This point was made by the 20th-century Austrian philosopher Paul Feyerabend in a paper published in 1960 (see the quote on the next page).

This is a subject American philosopher of science Norwood Russell Hanson explored in his book *Patterns of Discovery* (1958). Early on he gives the hypothetical example of Johannes Kepler and Tycho Brahe watching the same dawn. Kepler believed the Earth revolved around the Sun, while Brahe thought the opposite, that the Earth didn't move. So, 'do Kepler and Tycho see the same thing in the east at dawn?' Hanson asked. In terms of sense data, the answer must be an unequivocal 'yes'. But, Hanson argued, 'There is more to seeing than meets the eyeball'. Any observation of something is 'shaped by prior knowledge'. Kepler saw the dawn as the Earth spinning him back to face the Sun, while Brahe saw it as the beginning of the Sun's diurnal circuit around the Earth. As the fictional detective Sherlock Holmes critically put it, '... one begins to twist facts to suit theories, instead of theories to suit facts'. So, perhaps surprisingly, observed evidence isn't necessarily sufficient for supporting the truth of a theory.

FALSIFICATION

If there are difficulties proving the truth of a theory, then what other options does science have? Enter Austrian-British philosopher Karl Popper. In his 1934 book *Logik der Forschung* (*The Logic of Scientific Discovery*), he introduced the concept of falsificationism, which states that an attempt should always be made to prove a theory *wrong*. Now the first criterion for a scientific theory or hypothesis to be considered a good one is that it must be falsifiable. This, Popper insisted, was what distinguished science from what he considered to be pseudo-sciences such as Freudian and Marxist theories, which are virtually impossible to falsify (give it a try). The more falsifiable the theory, the better because, Popper felt, more was likely to be gleaned from any mistakes. There was little merit in playing it safe.

The idea of falsificationism is undoubtedly a useful one. But it has its limitations, illustrated, ironically, by one of Popper's favourite scientific examples. In 1919, English astronomer, physicist and mathematician Arthur Eddington sailed to an island off West Africa. The purpose of his journey was to test a prediction of Einstein's general theory of relativity regarding the bending of light by gravity through the observation of a solar eclipse. Eddington returned to declare the results as predicted by the theory. So far, so good – here seems the perfect demonstration of how falsificationism should work. We have a theory with big claims and had the results of the expedition been negative, then grave doubts would have appeared regarding it.

However, recent historians of science have looked carefully at the data Eddington collected and have concluded that he was in fact selective with his results. They certainly didn't vindicate Einstein's theory and could even have been presented as falsifying it. And yet the theory is still going strong today, precisely because of its excellent predictive power. The problem was that, at heart, the data Eddington recorded was poor – not good enough to prove or disprove the theory. What this shows is that a theory could be falsified through no fault of its own but because of a mistake elsewhere.

The announcement in 2011 of experimental results showing neutrinos travelling faster than light illustrates the issue further. At the time, reports focused on the challenge this presented to Einstein's special theory of relativity – the results, in effect, falsified it – and much was made of this. The hoo-ha, however, turned out to be due to a hardware error.

We interpret our
observation statements with the
help of the theories we possess …
and not the other way round.

**PAUL FEYERABEND (1924–1994)
IN 'PATTERNS OF DISCOVERY' IN**
THE PHILOSOPHICAL REVIEW
(1960)

Conceding to the Laws of Metre

In his poem 'The Vision of Sin' (1842),
Alfred, Lord Tennyson includes the lines:

> Every moment dies a man,
> Every moment one is born.

These lines prompted some wry editorial comments from
Charles Babbage, English mathematician and designer of the
analytical engine, arguably the first programmable computer.
Babbage wrote to the great man commenting, 'I need hardly
point out to you that this calculation would tend to keep the
sum total of the world's population in a state of perpetual
equipoise whereas it is a well-known fact that the said sum total
is constantly on the increase. I would therefore take the liberty
of suggesting that in the next edition of your excellent poem
the erroneous calculation to which I refer should be corrected
as follows:

> Every moment dies a man
> And one and a sixteenth is born.

I may add that the exact figures are 1.167, but something must,
of course, be conceded to the laws of metre.'

Needless to say, in the second edition of the poem,
published nine years later, Tennyson's lines stayed exactly as
they were.

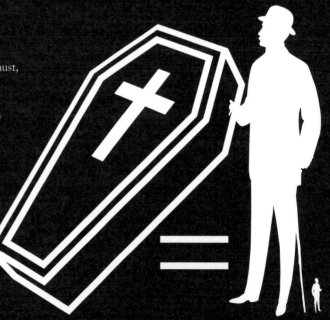

Machine and Man
Side by Side

THE THREE LAWS OF ROBOTICS

The Czech word *robota* refers to compulsory, or serf, labour. In his 1920 play *R. U. R. (Rossum's Universal Robots)*, the Czech writer and playwright Karel Čapek co-opted the word and coined 'robot' to designate a class of automatons designed to work for humans. In what has now become a standard turn of events for stories featuring such creations, they end up rebelling and destroying the human race.

Twenty-two years after the play was published, the science fiction author Isaac Asimov introduced 'the three laws of robotics' in his short story 'Runaround', which then formed part of his 1950 story collection *I, Robot*. The idea is that if followed, these laws should allow the safe coexistence of robots and human beings.

Asimov added a fourth law in later novels, which he named the Zeroth Law – 'zeroth' because it was meant to precede those laws above. Giving it this name also made the laws resonate with those of thermodynamics (see page 74).

1st LAW
A robot may not injure a human being, or, through inaction, allow a human being to come to harm.

2nd LAW
A robot must obey the orders given it by human beings except where such orders would conflict with the first law.

3rd LAW
A robot must protect its own existence as long as such protection does not conflict with the first or second Law.

An idea isn't responsible for the people who believe in it.

**DON MARQUIS,
'THE SUN DIAL COLUMN',
NEW YORK SUN, 1918**

ZEROTH LAW
A robot may not harm humanity, or, by inaction, allow humanity to come to harm.

THE ORIGINAL HIPPOCRATIC OATH

The *Hippocratic Corpus* is a collection of about sixty ancient Greek medical tracts written by a number of different authors over a fifty-year period during the 4th and 5th centuries BC. Included within it is the 'Hippocratic Oath', which lays down a physician's obligations. It's a document that has been rewritten throughout history to reflect the outlook of particular ages and cultures, which makes reading this 2002 translation of it in its original form by Michael North – then Head, Rare Books and Early Manuscripts at the US National Library of Medicine – all the more interesting.

*I swear by Apollo the physician, and Asclepius,
and Hygieia and Panacea and all the gods and goddesses
as my witnesses, that, according to my ability and judgement,
I will keep this Oath and this contract:*

*To hold him who taught me this art equally dear to me
as my parents, to be a partner in life with him, and to fulfil his
needs when required; to look upon his offspring as equals to my
own siblings, and to teach them this art, if they shall wish to learn
it, without fee or contract; and that by the set rules, lectures, and
every other mode of instruction, I will impart a knowledge of the
art to my own sons, and those of my teachers, and to students
bound by this contract and having sworn this Oath to the law of
medicine, but to no others.*

*I will use those dietary regimens which will benefit
my patients according to my greatest ability and judgement,
and I will do no harm or injustice to them.*

*I will not give a lethal drug to anyone if I am asked,
nor will I advise such a plan; and similarly I will not give a woman
a pessary to cause an abortion.*

*In purity and according to divine law will I carry out
my life and my art.*

*I will not use the knife, even upon those suffering from stones,
but I will leave this to those who are trained in this craft.*

*Into whatever homes I go, I will enter them for the benefit of
the sick, avoiding any voluntary act of impropriety or corruption,
including the seduction of women or men, whether they are
free men or slaves.*

*Whatever I see or hear in the lives of my patients, whether in
connection with my professional practice or not, which ought
not to be spoken of outside, I will keep secret, as considering
all such things to be private.*

*So long as I maintain this Oath faithfully and without corruption,
may it be granted to me to partake of life fully and the practice
of my art, gaining the respect of all men for all time. However,
should I transgress this Oath and violate it, may the
opposite be my fate.*

The Size and Gravity of the Situation

AN EARLY ATTEMPT TO CALCULATE EARTH'S CIRCUMFERENCE

Ancient Greek scientist Eratosthenes of Cyrene (c. 276–195 BC) was aware that at noon on the day of the summer solstice a rod planted in the ground at Syene (modern-day Aswan in Egypt) cast no shadow. This meant that the Sun was directly overhead. At exactly the same time, a rod at Alexandria cast a shadow showing the angle between the rod and the direction of the Sun's rays to be 1/50th of a circle. Using some knowledge of geometry, it's easy to see that this angle is equal to the angle between the two rods from the Earth's centre, or 7.2° (⅕₀th of 360°).

NOONDAY SHADOWS
The angle cast by the rod at Alexandria is equal to the angle between it and a rod planted in Syene.

A

A

Centre of
the Earth

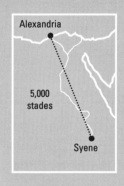

Alexandria

5,000
stades

Syene

All this experimentation and calculation means that the distance from Syene to Alexandria, which Eratosthenes was able to estimate at 5,000 stades (a Greek unit of measurement), was 1/50th of the Earth's circumference. A stade – which is one lap of a stadium – is thought by some scholars to be about 157.5m (517ft). If this was the length Eratosthenes had in mind, then that makes his calculation of the Earth's circumference more than 98 per cent accurate, given that the resultant figure is 39,375km, just 700km out from the true figure, which is 40,075km (24,901 miles). Even allowing for a more likely stade length of 185m (607ft) that is more generally accepted, which would give a circumference of 46,250km and an error of just over 15 per cent, Eratosthenes's achievement is still remarkable when we realise he lived at a time when it wasn't even clear that the Earth was spherical.

Of course, Eratosthenes's calculation assumes that the Earth is a perfect sphere, which it isn't. Like many of us, our planet exhibits an equatorial bulge and has been described as 'slightly potato shaped'. It's not of uniform density either. Gravitational pull at different locations on a non-uniform mass will vary through being closer or further from the centre of gravity of that mass. This is one of the major reasons why gravity varies by as much as ±0.5 per cent around the world – if you ever want to lose some weight, then one piece of advice would be to move to the equator. Sadly, while your weight would change, your mass would stay the same, so there would be little health benefit.

In another, more idiosyncratic, experiment, a garden gnome called Kern – in the enviable position of having a fixed mass – began blogging in late 2011 about his travels around the world. At various locations, he was weighed, using the same model of electronic scale each time. The idea was to investigate precisely how Kern's weight changed, depending where in the world he was. Here are some of his findings:

Location	Weight
The South Pole	309.82g
Under Newton's tree, Woolthorpe	308.59g
Balingen, Germany (Kern's hometown)	308.26g
Tokyo, Japan	307.9g
Sydney, Australia	307.8g
CERN, Geneva	307.65g
Mexico City, Mexico	307.62g
Mumbai, India	307.56g

Your Reading List for this Week Is ...

SETTLE DOWN WITH A GOOD SCIENCE BOOK

In 1999, the US publishing imprint Modern Library created a list of the 100 best non-fiction titles of the 20th century. It did not go uncriticised, particularly regarding the ranking of the books (many of the judges weren't aware this was happening when they made their suggestions), but included in the list were a number of science and mathematics titles, if you're looking for a good read (once you're finished with this book, obviously). The number in brackets at the end of each entry refers to the book's ranking in the complete list of 100.

Silent Spring
(1962), Rachel Carson (5)

The Double Helix
(1968), James D. Watson (7)

The Lives of a Cell
(1974), Lewis Thomas (11)

Principia Mathematica
(1910, 1912, 1913), Alfred North Whitehead and Bertrand Russell (23)

The Mismeasure of Man
(1981), Stephen Jay Gould (24)

The Art of the Soluble
(1967), Peter B. Medawar (26)

The Ants
(1990), Bert Hölldobler and Edward O. Wilson (27)

On Growth and Form
(1917), D'Arcy Thompson (34)

Ideas and Opinions
(1954), Albert Einstein (35)

The Making of the Atomic Bomb
(1987), Richard Rhodes (37)

Science and Civilization in China
(1954), Joseph Needham (40)

The Structure of Scientific Revolutions
(1962), Thomas S. Kuhn (69)

Florence Nightingale
(1950), Cecil Woodham-Smith (74)

A Mathematician's Apology
(1940), G. H. Hardy (87)

Six Easy Pieces
(1994), Richard P. Feynman (88)

The Taming of Chance
(1990), Ian Hacking (98)

There exist only two kinds of modern mathematics books: ones which you cannot read beyond the first page and ones which you cannot read beyond the first sentence.
CHEN NING YANG, NOBEL PRIZE WINNER 1957

All Together for a Tea Luncheon

A PARTY OF FAMOUS PHYSICISTS

Not dissimilar to the earlier 'How Do They Do IT' entry (see page 17), the following is another popular scientific joke.

One day, all of the world's famous physicists decided to get together for a tea luncheon. Fortunately, the doorman was a graduate student, and able to observe some of the guests …

- Everyone gravitated towards Newton, but he just kept moving around at a constant velocity and showed no reaction.
- Einstein thought it was a relatively good time.
- Coulomb got a real charge out of the whole thing.
- Cavendish wasn't invited, but he had the balls to show up anyway.
- Cauchy, being the only mathematician there, still managed to integrate well with everyone.
- Thomson enjoyed the plum pudding.
- Pauli came late, but was mostly excluded from things, so he split.
- Pascal was under too much pressure to enjoy himself.
- Ohm spent most of the time resisting Ampere's opinions on current events.

- Hamilton went to the buffet tables exactly once.
- Volt thought the social had a lot of potential.
- Hilbert was pretty spaced out for most of it.
- Heisenberg may or may not have been there.
- The Curies were there and just glowed the whole time.
- Van der Waals forced himself to mingle.

- Wien radiated a colourful personality.
- Millikan dropped his Italian oil dressing.
- De Broglie mostly just stood in the corner and waved.
- Hollerith liked the hole idea.
- Stefan and Boltzmann got into some hot debates.
- Everyone was attracted to Tesla's magnetic personality.
- Compton was a little scatter-brained at times.
- Bohr ate too much and got atomic ache.
- Watt turned out to be a powerful speaker.
- Hertz went back to the buffet table several times a minute.
- Faraday had quite a capacity for food.
- Oppenheimer got bombed.

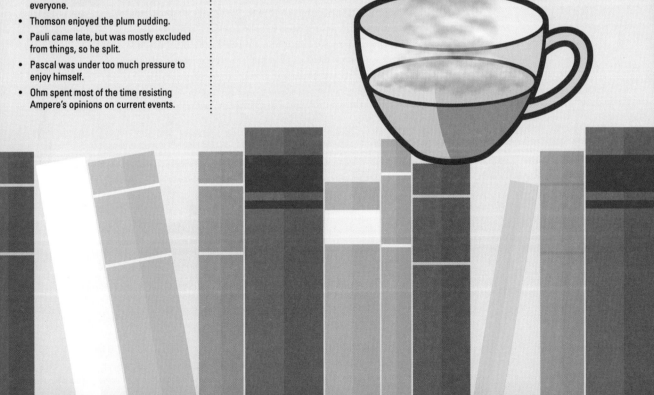

An Unacknowledged Debt

THE INFLUENCE OF ISLAMIC SCHOLARS

Western science owes a considerable debt to Islamic academics, particularly between the 8th and 15th centuries, a period commonly, if misleadingly, referred to as the Dark Ages. It's a time traditionally seen as a low point in the advancement of knowledge, after all the terribly clever things the Greeks and Romans had been up to before that – but nothing could be further from the truth.

In this period, the Persian astronomer and mathematician Mūsā al-Khwārizmī, born around 786, developed algebra. It was also during this time that the great 'translation movement' took place, a huge intellectual project that saw a mass of texts in ancient Greek and other languages translated into Arabic, which later enabled these ideas to flow back into Europe. One of these translations involved the work of physician Ibn Sina (his Latinised name is Avicenna), who around the turn of the second millennium wrote his multi-volume *Al-Qanun Fi Al-Tibb* (*The Canon of Medicine*), which became one of the standard medical textbooks in Europe until the late 17th century.

The appearance of movement in a straight line when a circle rolls inside another that's twice its radius.

Scholars today are constantly discovering new connections between the Islamic world of the Dark Ages and some of the 'breakthroughs' in science that took place elsewhere just afterwards. One of the most interesting of these has been the gradual understanding that Polish astronomer Nicolaus Copernicus probably used mathematical models first developed by Islamic astronomers in his proposal that the Earth revolved around the Sun.

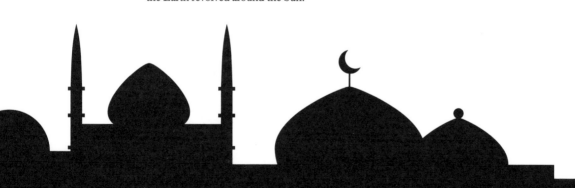

Copernicus did acknowledge a debt to Islamic thinking in his groundbreaking book *De revolutionibus orbium coelestium* (*On the Revolutions of the Heavenly Spheres*), published in 1543, although there was nothing specifically mentioned that dates to later than around 1200. However, we now know that within Islamic astronomy there was already a wealth of mathematically based criticism of the Ptolemaic view (with the Earth at the centre of the solar system) that Copernicus's system was effectively overthrowing. As there was no tradition in Europe of questioning Ptolemy, the conviction is now growing that some of the Islamic work directly influenced Copernicus.

F. Jamil Regep, a professor at McGill University in Canada, has provided the following tangible examples of possible Islamic influence:

- Probable use of the Tusi couple, a method developed by the Persian scholar Naṣīr al-Dīn al-Ṭūsī during the 13th century to show how the appearance of movement in a straight line can be produced by a circle rolling inside another circle twice its radius. (It's worth watching one of the animations of the Tusi couple on the internet to see how it works.) This method enabled Copernicus to deftly sidestep a feature of Ptolemy's system. The lettering of a relevant diagram in *De revolutionibus* follows the customary Arabic form rather than Latin, despite the book being written in the latter language.

- There's a striking similarity between a section of *De revolutionibus* and one in al-Tusi's book *Tadhkirah*, which appeared in many subsequent Islamic works.

- There are astronomical models featured in the work of Ibn al-Shāṭir, who worked in Damascus in the 14th century, which, until their discovery, had been previously thought to have originated with Copernicus. The implication is that Copernicus had seen them and used them in his work.

Elements of Colour

TITANIUM
Intense white

ALUMINIUM
Intense white

THE RAINBOW SCIENCE OF FIREWORKS

The characteristic colours of fireworks are due to metals, and their salts, being added to the gunpowder. Particular hues are due to particular metals and their salts – exactly the same ones are seen in laboratory flame tests.

When a metal salt is heated, such as in a flame or the ignited gunpowder, the electrons in the metal atoms get promoted to higher energy levels. When the electrons return to their normal, or ground, state, energy is emitted as light at wavelengths specific to that metal. In the case of our fireworks, these frequencies lie in the visible region, and it's the frequency that determines the colour we then see.

SODIUM
Yellow

CALCIUM
Orange-red

LITHIUM
Red

COPPER
Blue

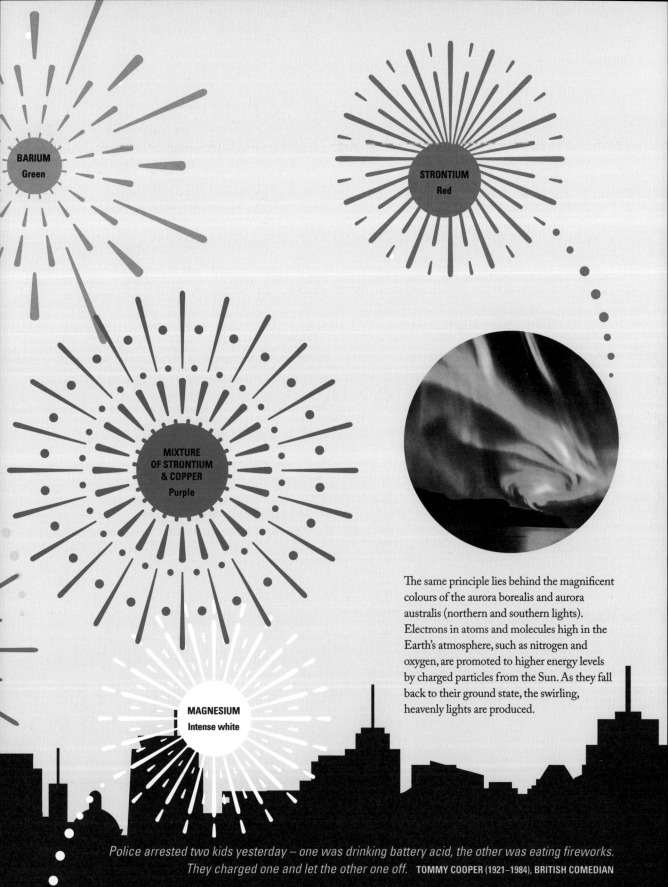

BARIUM
Green

STRONTIUM
Red

MIXTURE OF STRONTIUM & COPPER
Purple

MAGNESIUM
Intense white

The same principle lies behind the magnificent colours of the aurora borealis and aurora australis (northern and southern lights). Electrons in atoms and molecules high in the Earth's atmosphere, such as nitrogen and oxygen, are promoted to higher energy levels by charged particles from the Sun. As they fall back to their ground state, the swirling, heavenly lights are produced.

Police arrested two kids yesterday – one was drinking battery acid, the other was eating fireworks. They charged one and let the other one off. **TOMMY COOPER (1921–1984), BRITISH COMEDIAN**

The Ten Greatest Ever Equations
... According to Nicaragua

STAMPS OF MATHEMATICAL APPROVAL

In 1971, the Nicaraguan postal service issued a series of postage stamps depicting the ten mathematical equations that, in its opinion, had most changed the world. Each stamp included on its reverse a justification of its selection. Here they are, translated:

$1 + 1 = 2$

PRIMITIVE MAN

Simple as it is, this equation had enormous consequences for humanity because it formed the basis of counting. Without an understanding of numbers, people could only trade in a rudimentary way; they didn't have an exact tally of the number of sheep or cows or how many men made up the tribe. The discovery led directly to the rapid development of trade and later to the important science of measurement.

$a^2 + b^2 = c^2$

PYTHAGORAS (570–497 BC)

The most frequently used theorem in geometry is undoubtedly that of Pythagoras, which refers to the lengths of the three sides – a, b and c – of a right-angle triangle. It provided for the first time a means of calculating lengths by indirect means, thereby allowing man to do surveying and produce maps. The ancient Greeks used it to measure such things as the distances of ships at sea and the heights of buildings. Today, scientists and mathematicians use it constantly.

$$F_1{}^{x_1} = F_2{}^{x_2}$$

ARCHIMEDES (281–212 BC)

Archimedes said, 'Give me a place to stand and I will move the world'. The simple equation of the lever is the basis of all engineering, whether it be merely with a crowbar, or the most advanced gearing or crane. It's essential for the design of machinery and all structures from bridges to buildings. Every nut and bolt employs the principle. Our car brakes, door handles, scales and most tools are varieties of the lever.

$$e^{\ln N} = N$$

JOHN NAPIER (1550–1617)

With the invention of logarithms, Napier gave the world a powerful shorthand for arithmetic. It enabled men to do multiplication or division by simply adding or subtracting the logarithms of numbers, which allowed them to carry these operations out faster, as well as more complicated ones containing many numbers. The impact of logarithms in fields such as astronomy and navigation was huge and comparable to the modern era's computer revolution.

$$f = \frac{Gm_1m_2}{r^2}$$

ISAAC NEWTON (1642–1727)

Before Newton's time, people had little idea of the force that kept the planets in their orbits around the Sun, or the Moon around the Earth – or even what stopped us from flying off the surface of the Earth into space. Newton showed that the force of gravity attracts all bodies to one another. The equation shows, however, that although the force depends on the masses of the bodies, we do not notice it between everyday objects because it's comparatively very weak.

$$\nabla^2 E = \frac{Ku}{c^2} \frac{\delta^2 E}{\delta t^2}$$

JAMES CLERK MAXWELL (1831–1879)

A century ago, this Scottish physicist discovered four famous equations summarising man's knowledge of electricity and magnetism. From these he then obtained this one equation, along with another predicting the possibility of radio waves. We owe Maxwell all our TV and radio broadcasting, our long-distance communication and radar on land, at sea and in space. Light, X-rays and other electromagnetic radiation are also governed by this fundamental equation.

$$S = k \log W$$

LUDWIG BOLTZMANN (1844–1906)

Boltzmann's equation revealed how the behaviour of gases depended on the constant motion of atoms and molecules. Its great importance lies in its application in areas in which gases play an important role: in all machines driven by steam or internal combustion; in gas reactions of chemicals used to make modern drugs, plastics or other substances; in understanding weather; and even to explain the violent processes of the Sun, stars and distant galaxies.

$$v = v_e \ln \frac{m_0}{m_1}$$

KONSTANTIN TSIOLKOVSKY (1857–1935)

A basic part of space technology, this equation gives the changing velocity of a ship as it burns the fuels it's carrying. The equation is derived directly from one of the three laws of motion of Isaac Newton. Without it, launching spacecraft to the Moon and planets or orbiting the Earth would be impossible, and it has also made practicable the use of guided rockets in warfare.

*It is of the highest importance
in the art of detection to be able to
recognise out of a number of facts
which are incidental and which vital.*

**SHERLOCK HOLMES IN
'THE ADVENTURE OF THE REIGATE PUZZLE',
BY ARTHUR CONAN DOYLE
(1859–1930)**

$E = mc^2$

ALBERT EINSTEIN (1879–1955)

This equation lies at the root of our nuclear age. It simply
says that a small quantity of matter can be converted into a
large amount of energy. We see this nuclear energy released
in a spectacular and violent way in atomic and hydrogen
bombs. But man has also managed to tame 'nuclear fission'
in reactors that supply heat and generate electricity for our
homes and factories.

$\lambda = {}^h/_{mv}$

LOUIS DE BROGLIE (1892–1987)

Light, a transfer of energy, can behave both as a particle and
as a continuous wave. De Broglie discovered the converse,
that the elementary particles of which matter is composed
also have properties that resemble waves. His equation has
had a large effect on physics, leading to modern optics
(photons), electronic components (e.g. transistors) with
many applications in radio, TV, computers, spacecraft and
military weapons. It also provided scientists with the
powerful electron microscope.

Appendix

DOUBLE SCIENCE

School science – did it fill you with dread or delight? For me, that rather depended on the teacher and the area being covered. (I'll let you in on a secret; I never got on with electric circuits, which is probably why you won't find anything to do with them in this book.) But school science is the grounding of most, if not all, of our scientific literacy, so it doesn't do any harm to go over some of the things you may have covered there.

We'll start with one of my favourites (don't get cornered by me at a party – seriously), the atom.

CHEMISTRY IS FULL OF BAD JOKES BECAUSE ALL THE BEST ARGON

If you search the depths of your memory, you may recall that the periodic table contains the 118 elements from which *everything* is made and that the smallest unit of any element is the atom. Atoms themselves are made up of three subatomic particles: protons, neutrons and electrons (you'll discover in the 'Making the Standard Model' entry – see pages 86–89 – that it's actually a whole lot more involved than this, but let's not overcomplicate things for now). The nucleus of an atom, found at its centre, is made up of the neutrons and protons clustered together. Only these contribute to the element's atomic mass. Both neutrons and protons have

roughly the same mass. The electrons, on the other hand, have a negligible mass in comparison to the nucleus, which they 'orbit'.

The elements in the periodic table are arranged in order of increasing atomic number, which is determined by the number of protons – all the atoms of any given element always have the same number of protons. The atomic number of carbon, for example, is 6, which means there are six protons in every atom of carbon.

While every atom of an element has the same number of protons, the same isn't always true of neutrons. Isotopes are atoms of the same element that differ in the number of neutrons they contain. For example, there are three naturally occurring isotopes of carbon: carbon-12, carbon-13 and carbon-14 (the number after the dash refers to the atomic mass of the isotope; so, given that carbon always contains six protons, carbon-13 must contain seven neutrons).

As we just saw, electrons orbit the nucleus of an atom, where its protons and neutrons are packed together. Protons have a relative charge of +1 and electrons have a relative charge of –1 (neutrons are uncharged). So, having an equal number of protons and electrons means that the atom is electrically neutral overall. For example, a neutrally charged atom of carbon will contain six protons – hence its atomic number is 6 – and six electrons.

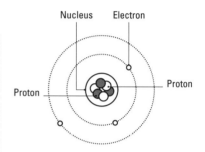

The structure of an atom

Electrons orbiting a nucleus do so according to clearly defined regions called electron shells. Each shell, which corresponds to a level of electron energy, can contain only a fixed number of electrons. The first one, which is the lowest-energy shell, can hold a maximum of two electrons only. Each subsequent shell of the first twenty elements of the periodic table can contain eight electrons. As you move along the periodic table in order of increasing atomic number, these shells are filled up with electrons in sequence. So, hydrogen (the first element in the periodic table) has one electron in the first shell and helium has two. Now the first shell is full, the next element, lithium, will have two electrons in the first shell and one electron in the second shell. Once we get to neon (with the atomic number 10), both the first shell (two electrons) and second shell (eight electrons) are now full, so the next element, sodium (with atomic number 11), has a single electron in the third shell. The outermost shell of an atom is called the valence shell.

If an atom somehow loses or gains an electron, meaning that there are more protons than electrons, or vice versa, it becomes an ion. If an electron is lost, a positively charged ion (cation) is formed; if an electron is gained, a negatively charged ion (anion) is formed. Metals will generally lose electrons to form cations while non-metals will gain electrons to form anions. This can happen when metals and non-metals react with each other, such as when sodium and chlorine combine to form sodium chloride (common table salt). Here, a chlorine atom effectively takes an electron from a sodium atom, and an ionic bond is formed between them. One useful way of remembering the difference between cations and anions is the phrase 'cations are pussy-tive'. A chemical ion is denoted by adding the relevant charge after the element's symbol, in superscript – for example: $Li+$, Ca^{2+}, $F-$ and O^{2-}.

CELL-U-LIKE

So, all matter is made of one or more of the elements of the periodic table. On a larger (but still microscopic) scale, all living things, called organisms, are made up of cells. Animal and plant cells are similar in many ways but, as the illustration here shows, there are also some fundamental differences.

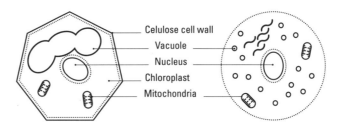

Celulose cell wall
Vacuole
Nucleus
Chloroplast
Mitochondria

A plant cell (left) and an animal cell (right)

The features common to both types of cell are:

- The cell membrane is a very thin, porous wall that enables necessary gases and chemical nutrients to pass into and out of the cell.

- The nucleus is effectively the 'control room' of the cell. It also contains the cell's genetic material.

- Both cell types possess a fluid environment outside the nucleus in which everything else in the cell resides; it's called cytoplasm.

- In plant cells there are large, sac-like areas in the cell called vacuoles. They contain high concentrations of nutrients and assist in the disposal of waste. Not all animal cells possess vacuoles, but in those that do they're much smaller in size and more numerous than in plant cells.

- Mitochondria are the 'powerhouses' of both cell types, since they produce the molecule adenosine triphosphate (ATP), the currency of energy in all living organisms.

The differences between plant and animals cells are significant, however:

- Only the plant cell contains a rigid, cellulose wall. This is what gives plants their great strength.

- Only plant cells contain chloroplasts – structures which in turn contain chlorophyll. They give plants their green colour and are responsible for photosynthesis, the method by which plants turn carbon dioxide and water into sugars such as glucose to provide themselves with energy.

Appendix

WHO DO YOU THINK YOU ARE?

The concept of the gene, the basic unit of inheritance, has been understood for more than a century. But it wasn't until molecular biologists James Watson and Francis Crick – with important contributions from chemist Rosalind Franklin – cracked the structure of deoxyribonucleic acid (DNA) that the study of genetics really began to take off.

DNA is the molecule in which all our genes are located and it can be found in all our bodies (a gene is a section of DNA – they vary in size). The now-famous structure of DNA is a double helix – a little like a ladder twisted into a spiral (above).

Although the information in the DNA of an organism determines most of what that thing is, it can't predict everything precisely down to the last detail. It's probably better to think of DNA as a recipe or even a script to be interpreted rather than taking the common view of it being a blueprint.

Eight hundred lifespans can bridge more than 50,000 years. But of these 800 people, 650 spent their lives in caves or worse; only the last 70 had any truly effective means of communicating with one another, only the last 6 ever saw a printed word or had any real means of measuring heat or cold, only the last 4 could measure time with any precision; only the last 2 used an electric

The DNA double helix

motor; and the vast majority of the items that make up our material world were developed within the lifespan of the 800th person.

'Assessing Technology Transfer' (NASA Report SP-5067), 1966, pages 9–10

In both plants and animals, DNA is packaged up (along with other materials) in the form of chromosomes, which look like two thread-like structures pinched together at a point along their length. There are forty-six chromosomes in twenty-three pairs in every cell of the human body. One of these pairs will be made of two X chromosomes if you're female and an X and Y chromosome if you're male (only men carry a Y chromosome). The disorder Down's syndrome, for instance, is due to a person having an extra copy of chromosome 21.

When sperm and egg cells are formed, only one chromosome from each pair ends up being in each sperm or egg cell. So, in every egg cell there will be an X chromosome (either one from the female's XX pair), whereas half of the sperm cells will contain an X chromosome and the other half will contain a Y one (either one from the male's XY pair). When a sperm and egg cell fuse, the number of chromosomes is restored to forty-six, with half of the genetic material coming from the father and the other half from the mother. Whether the sperm cell involved contains an X or a Y chromosome will ultimately determine the baby's sex.

BY A FORCE OF NATURE

Isaac Newton (1642–1727) is famous for many things, but at school you're most likely to have come across his three laws. This is how Newton himself expressed them:

FIRST LAW

Every body perseveres in its state of rest, or of uniform motion in a right line, unless it is compelled to change that state by forces impressed thereon.

The first part of this statement introduces the concept of inertia. The second part goes on to tell us this property of matter is only affected by an external force. So, a body will stay in a state of rest or will continue to keep moving if already doing so unless an external force is applied. If you imagine there being a toy car sitting on your table, this remains stationary until you apply a force on it – by pushing it, for example. As soon as you do push it, other forces start to act against it, the most notable being friction. The car would travel a lot further if pushed in space because there would be fewer and weaker forces subsequently acting against it.

SECOND LAW

The alteration of motion is ever proportional to the motive force impressed; and is made in the direction of the right line in which that force is impressed.

The second law is best summarised by the equation $F = ma$, or force equals mass times acceleration. It's clear from this equation that the more force you apply, the greater the subsequent acceleration. Imagine kicking a football with all your might compared with kicking it gently. The equation also tells us that the greater the mass something has, the greater the force that needs to be applied to get it to move. Imagine the difference between pushing a toy car and pushing a real one.

This brings us to the very important difference between mass and weight. Mass is measured in kilograms (kg) and is about how much stuff of something there is. It's the same wherever you are in the universe.

However, the weight of something is dependent on gravity and is measured in Newtons (N), which is the unit of force. So, based on our equation above, weight of an object with mass, m, corresponds to the equation $W = ma$. Here 'a' corresponds to gravitational acceleration, g, and the equation in this instance becomes $W = mg$. On Earth, g has the value 9.8 m/s². On the Moon, it has the value 1.63 m/s². So, a person with a mass of 70kg would weigh 686 N on the Earth but on the Moon would weigh about a sixth of this. So, while the mass of something cannot change, the weight very much can.

THIRD LAW

To every Action there is always opposed an equal Reaction: or the mutual actions of two bodies upon each other are always equal, and directed to contrary parts.

It's no great revelation when I say the Earth's gravity is acting upon you all the time. It's what keeps our feet on the ground, after all. But Newton's third law tells us that an equal (and opposing) force must also be happening. With this law, more than one object has to be involved. In the case of someone standing still, the ground is what is applying this. Another good example is a gun being fired. The bullet shoots out at great speed and, in return, the gun recoils. To prevent the gun from physically moving, an equal force has to be applied by the person firing the gun.

Blind commitment to a theory is not an intellectual virtue: it is an intellectual crime.

Imre Lakatos (1922–1974), Hungarian philosopher of science

Selected Bibliography

Oxford Dictionary of Science, Oxford University Press, Oxford, 2010

'University of Cambridge Local Examinations Syndicate question papers, timetable, grace, regulations and notices, 1858'

'University of Cambridge Local Examinations Syndicate First Annual Report, 1859'

Nineteenth-Century Women Poets. Oxford University Press: Oxford, 1996.

Benfey, Theodor. 'The Biography of a Periodic Snail'. *Bulletin of the History of Chemistry*, Vol. 34, No. 2.

Bizony, Piers. *Atom.* Icon Books: London, 2007.

Brock, Claire. *The Comet Sweeper.* Icon Books: London, 2007.

Bruce, Susan. *Three Early Modern Utopias.* Oxford University Press: Oxford, 1999.

Bynum, W. F., and Roy Porter. *Oxford Dictionary of Scientific Quotations.* Oxford University Press: Oxford, 2005.

Campbell, Lewis and William Garnett. *The Life of James Clerk Maxwell.* Macmillan and Co.: London, 1882.

Carey, Nessa. *The Epigenetics Revolution.* Icon Books: London, 2011.

Chalmers, A. F. *What is this thing called Science?.* Open University Press: Milton Keynes, 1999.

Cheater, Christine. 'Collectors of Nature's Curiosities'. *Frankenstein's Science.* Ashgate: Aldershot, 2008.

Clegg, Brian. *Inflight Science.* Icon Books: London, 2011

Clerk Maxwell, James. *Theory of Heat.* Longmans, Green, and Co.: London, 1902.

Cruciani, Fulvio, Beniamino Trombetta, Andrea Massaia, Giovanni Destro-Bisol, Daniele Sellitto and Rosaria Scozzari. 'A Revised Root for the Human Y Chromosomal Phylogenetic Tree: The Origin of Patrilineal Diversity in Africa'. *American Journal of Human Genetics*, Vol. 88, No. 6 (2011).

Darwin, Charles. *A Naturalist's Voyage Round the World.* John Murray: London, 1913.

Darwin, Charles. *On the Origin of Species.* Penguin Books: London, 2009.

Davies, B. 'Edme Mariotte 1610–1684'. *Physics Education*, Vol. 9, No. 275 (1974).

Ellegård, Alvar. *Darwin and the General Reader: 1859–1872.* Göteborg: Stockholm, 1958.

Elsdon-Baker, Fern. *The Selfish Genius.* Icon Books: London, 2009.

Engels, David. 'The Length of Eratosthenes's Stade'. *American Journal of Philology*, Vol. 106, No. 3 (Autumn, 1985).

Fara, Patricia. *Scientists Anonymous.* Wizard Books: London, 2005.

Feynman, Richard. *Six Easy Pieces.* Penguin Books: London, 1995.

Flood, W. E. *Scientific Words.* The Scientific Book Guild. 1961.

Graham, Jr, C. D. 'A Glossary for Research Reports'. *Metal Progress*, Vol. 71, No. 5 (1957).

Gregory, Andrew. *Eureka.* Icon Books: London, 2001.

Gullberg, Jan. *Mathematics: From the Birth of Numbers.* W. W. Norton & Company: New York, 1997.

Haile, N. S. 'Preparing Scientific Papers'. *Nature,* Vol. 268 (1977).

Hannam, James. *God's Philosophers.* Icon Books: London, 2009.

Hanson, Norwood Russell. *Patterns of Discovery.* Cambridge University Press: Cambridge, 1965.

Haynes, Clare. 'A Natural Exhibitioner: Sir Ashton Lever and his Holophusikon'. *British Journal for Eighteenth-Century Studies*, 24 (2001).

Hazen, Robert M. and James Trefil. *Science Matters.* Anchor Books: New York, 2009.

Henry, John. *Knowledge is Power.* Icon Books: London, 2002.

Henry, John. Moving *Heaven and Earth.* Icon Books: London, 2001.

Hogg, Thomas Jefferson. 'Shelley at Oxford'. *New Monthly Magazine* (1832).

Huxley, Leonard. *Life and Letters of Thomas Henry Huxley* – Volume 1. MacMillan and Co.: London, 1903.

James, Arthur M. and Mary P. Lord. *Macmillan's Chemical and Physical Data.* The Macmillan Press: London, 1992.

Jones, Bence. *The Life and Letters of Faraday*. Longmans, Green, and Co.: London, 1870.

Kipperman, Mark. 'Coleridge, Shelley, Davy, and Science's Millennium'. *Criticism*, Vol. 40, No. 3 (1998).

Summer Knight, David. 'Humphry Davy the Poet'. *Interdisciplinary Science Reviews*. Vol. 30, No. 4 (2005).

Knowles, Elizabeth, ed. *The Oxford Dictionary of Quotations*. Oxford University Press: Oxford, 1999.

Kuhn, Thomas S. *The Structure of Scientific Revolutions*. University of Chicago Press: Chicago, 2012.

Kumar, Manjit. *Quantum*. Icon Books: London, 2008.

Ladd, Everett Carll, Jr. and Seymour Martin Lipset. 'Politics of Academic Natural Scientists and Engineers'. *Science*, Vol. 176, No. 4039 (1972).

Lamont-Brown, Raymond. *Humphry Davy*. Sutton Publishing: Stroud, 2004.

Mackay, Alan L. *A Dictionary of Scientific Quotations*. IOP Publishing: Bristol, 1991.

Masood, Ehsan. *Science & Islam*. Icon Books: London, 2009.

Mpemba, E. B. and D. G. Osborne. 'Cool?'. *Physics Education*, Vol 4 (1969).

Newton, Isaac. *The Mathematical Principles of Natural Philosophy*, Volume 1. Benjamin Motte: London, 1729.

OECD. *Science, Technology and Industry Outlook*. Page 43 (2010).

Pickover, Clifford A. *A Passion for Mathematics*. John Wiley & Sons: Hoboken, 2005.

Poe, Edgar Allen. *Eureka*: A Prose Poem. Geo. P. Putnam: New York, 1848.

Pray, L. 'Eukaryotic Genome Complexity'. *Nature Education*, Vol. 1, 1 (2008).

Ragep, F. Jamil. 'Copernicus and his Islamic Predecessors'. *History of Science*, xlv (2007).

Rhys Morus, Iwan. *Michael Faraday and the Electrical Century*. Icon Books: London, 2004.

Ruston, S. *The Science of Life and Death in Frankenstein*. Bodleian Library Press: Oxford, 2021.

Sampson, F. *In search of Mary Shelley: The girl who wrote Frankenstein*. Profile Books: London, 2018.

Schrödinger, Erwin. 'The Present Situation in Quantum Mechanics'. Translated by Trimmer, John D. in Proceedings of the American Philosophical Society, Vol. 124, No. 5 (June 1980).

Secord, Jim. 'Newton in the Nursery: Tom Telescope and the Philosophy of Tops and Balls, 1761–1838'. *History of Science*, Vol. 23 (1985).

Snow, C. P. *The Two Cultures*. University of Cambridge: Cambridge, 1993.

Tange, Andrea Kaston. 'Constance Naden and the Erotics of Evolution: Mating the Woman of Letters with the Man of Science'. *Nineteenth-Century Literature*, Vol. 61, No. 2 (September 2006).

Tennent, R. M., ed. *Science Data Book*. Oliver & Boyd: Edinburgh, 1971.

Timbs, John. *The Year Book of Facts in Science and Arts*, 1855. David Bogue: London, 1855.

Toyabe, S., T. Sagawa, M. Ueda, E. Muneyuki and M. Sano. 'Experimental demonstration of information-to-energy conversion and validation of the generalized Jarzynski equality'. *Nature Physics*, Vol. 6, No. 12 (2010).

Waller, John. *The Discovery of the Germ*. Icon Books: London, 2002.

Weber, R. L. and E. Mendoza eds. *A Random Walk in Science*, IOP Publishing: London, 1973.

Weber, R. L. ed. *Droll Science*. Humana Press: Clifton, 1987.

Weber, R. L. ed. *More Random Walks in Science*. IOP Publishing: London, 1982.

Weber, R. L. ed. *Science with a Smile*. IOP Publishing: London, 1992.

White, William. *The Illustrated Hand Book of the Royal Panopticon of Science and Art*. John Hotson: London, 1854.

Index

Index

Picture Credits & Permissions

9 (all Shutterstock): © NASA; © Zonda; © Morphart Creation; © LineTale 11 (all Shutterstock): Car © KUNJAY ART; Goat © FRIsky photo 12: Rings © Christos Georghiou | Shutterstock 13 (all Shutterstock): Moon © Lina_Lisichka; Scales © Koshevnyk 14 (all Shutterstock): Glucose © Paper Teo; Cell © shopplaywood 15 (all Shutterstock): Dollar bill © Photocreo Michal Bednarek; Golf ball © Dario Sabljak; Butterfly © Anastasia_Panchenko; Football © Volonoff 16–17: © Jason Winter | Shutterstock 21: © Keith Corrigan | Alamy Stock Photo 22: © IanDagnall Computing | Alamy Stock Photo 24: © zeber | Shutterstock 27: © ledokolua | Shutterstock 28: © goodwin_x | Shutterstock 32: © Shmitt Maria | Shutterstock 34: © unterwegs | Shutterstock 36: © onot | Shutterstock 41: © Fred Espenak, NASA/GSFC 42: Telescope © Morphart Creation | Shutterstock 43: © helenpyzhova | Shutterstock 47: © Moab Republic | Shutterstock 52: © Wellcome Images, a website operated by Wellcome Trust, a global charitable foundation based in the United Kingdom 53: © Marzolino | Shutterstock 56: © Aha-Soft | Shutterstock 57: Volcano © bamban heru cahyanto | Shutterstock 66: © tristan tan | Shutterstock 68–69: © Business stock | Shutterstock 71: Pig © ussr | Shutterstock 72–73: © ArinaCo | Shutterstock 76–77: Cubes © MchlSkhrv | Shutterstock 78: Glass © Rainbow Black | Shutterstock 81: Column © Rvector | Shutterstock 84: © logo vectory | Shutterstock 85 (all Shutterstock): Savannah © Seita; Microbe © In-Finity 87: © allaboutvector | Shutterstock 92: © Morphart Creation | Shutterstock 94: © Master_Andrii | Shutterstock 97: © Niphon Subsri | Shutterstock 98: © Artokoloro 100: © Royal Astronomical Society; Science Photo Library' | Alamy Stock Photo 104 (all Shutterstock): Top © Bjoern Wylezich; Middle © Albert Russ; Lower © Mara Fribus 106 (all Shutterstock): Aircraft © researcher97; Teapot & saucer © Ilya Akinshin 114–115: © Vector Tradition | Shutterstock 121: © Science source | Science Photo Library 122–123: © tinkivinki | Shutterstock 125: © Liu zishan | Shutterstock 126–127: © helenpyzhova | Shutterstock 130 (all Shutterstock): Cricket pitch © Dileep; Mountains © SaveJungle; Eiffel tower © Nikolai V Titov; Moon © Lina_Lisichka 134–135: Lanterns © rycw; Background © Valeriya_Dor 137: © Polina Kudelkina | Shutterstock 138: © grecosvet | Getty Images 139 (clockwise from top left): © acrogame | Adobe Stock; PD; © Chronicle | Alamy Stock Photo; Shutterstock; © UniPress; © I000s_pixels; PD; © Trinity Mirror | Mirrorpix | Alamy Stock Photo; © UniPress; PD 143: © Pachim | Shutterstock 150–151: © Alhovik | Shutterstock 158 (all Shutterstock): Upper © saran insawat; Lower © Zern Liew 160: © Everett Collection |

Shutterstock 164: Background dots © Login | Shutterstock 169: AC generator © Hein Nouwens | Shutterstock 170: Middle left © Mailsapartbassimore | CC BY-SA 4.0 171: Upper right © Keystone | Getty Images; Middle right © Leslie Carolyn Edwards; Lower left © Lizzie Caswall Smith; Lower right © Henry Grant Collection/Museum of London 175: © Marzolino | Shutterstock 179: Pea blossoms © Dunaeva Natalia | Shutterstock 182 (all Shutterstock): Cherries © Slonomysh; © Marzufello 183: Skull © Betacam-SP | Shutterstock; Pie © Morphart Creation | Shutterstock; Newton's tomb © Alamy Stock 185 (all Shutterstock): Upper © Morphart Creation; Upper right © Hein Nouwens 187 (all Shutterstock): Human figures © majivecka; Earth © mejnak 192: Flag © Volonoff | Shutterstock 193: Gold © Rvector | Shutterstock 198 (all Shutterstock): Coffin © Aayam 4D; Person © alex74 200: Wellcome Images, a website operated by Wellcome Trust, a global charitable foundation based in the United Kingdom 203 (all Shutterstock): Upper © researcher97; Lower © BlueRingMedia 209: Aurora © Mike-Hubert.com | Shutterstock 210–213: © Correos de Nicaragua 216: © Vladgrin | Shutterstock

Every effort has been made to trace copyright holders and acknowledge the images used in this book. The publishers welcome further information regarding an unintentional omissions.

Hans Van de Bovenkamp, executor for the Siv Cedering Estate, for allowing the reproduction of the poem , 'Letter from Caroline Herschel' (1750, 1848).

Theodor Benfey for allowing his periodic snail to be redrawn and reproduced.

'Preparing scientific papers', reprinted by permission from Macmillan Publishers Ltd: Nature 268, 1977.

Professor Paul May, for using material from his silly molecules website: www.chm.bris.ac.uk/sillymolecules/sillymols.htm.

'Glossary for Research Reports', reprinted with permission of ASM International. All rights reserved. www.asminternational.org.

Professor Mike Benton for allowing his spindle diagram in *Vertebrate Palaeontology* to be redrawn and reproduced.

Kate McAlpine (www.katemcalpine.com) for allowing her LHC lyrics to be reproduced.

'The Tale of Schrödinger's Cat', by Marilyn T. Kocher reprinted with permission from Physics Today, May 1978, American Institute of Physics.

This illustrated edition published in the UK in 2022 by Icon Books Ltd, Omnibus Business Centre, 39–41 North Road, London N7 9DP
info@iconbooks.com
www.iconbooks.com

Previously published in the UK in 2012 by Icon Books Ltd

Sold in the UK, Europe and Asia by Faber & Faber Ltd, Bloomsbury House, 74-77 Great Russell Street, London WC1B 3DA or their agents

Distributed in the UK, Europe and Asia by Grantham Book Services, Trent Road, Grantham NG31 7XQ

Distributed in Australia and New Zealand by Allen & Unwin Pty Ltd, PO Box 8500, 83 Alexander Street, Crows Nest, NSW 2065

Distributed in South Africa by Jonathan Ball, Office B4, The District, 41 Sir Lowry Road, Woodstock 7925

Distributed in India by Penguin Books India, 7th Floor, Infinity Tower-C, DLF Cyber City, Gurgaon 122002, Haryana

ISBN: 978-1-78578872-7

Text copyright © Simon Flynn, Design copyright © UniPress Books Ltd 2022

The author asserts his moral rights.

All rights reserved. No part of this book may be reproduced, transmitted or stored in an information retrieval system in any form or by any means, graphic, electronic or mechanical, including photocopying, taping and recording, without prior written permission from the publisher.

DESIGN AND ART DIRECTION:
Wayne Blades

Printed in China

1 2 3 4 5 6 7 8 9 10